High-Stakes Aviation:
U.S.-Japan Technology Linkages in
Transport Aircraft

Committee on Japan
Office of Japan Affairs
Office of International Affairs
National Research Council

National Academy Press
Washington, D.C. 1994

This project was made possible with funding support from the U.S. Department of Defense, the U.S. Department of Commerce, and the Japan-United States Friendship Commission.
Available from:

National Academy Press
2101 Constitution Avenue, N.W., Box 285
Washington, D.C. 20055
800-624-6242 or 202-334-3313 (in the Washington Metropolitan Area).

Library of Congress Catalog Card Number 94-65759
International Standard Book Number 0-309-05045-6
B-322

COMMITTEE ON JAPAN

Erich Bloch, *Chairman*
Council on Competitiveness

Richard J. Samuels, *Vice-Chairman*
Massachusetts Institute of Technology

Sherwood L. Boehlert
U.S. House of Representatives

John O. Haley
University of Washington

Lewis M. Branscomb
Harvard University

Jim F. Martin
Rockwell International

G. Steven Burrill
Burrill & Craves

Joseph A. Massey
Dartmouth College

Lawrence W. Clarkson
The Boeing Co.

Mike M. Mochizuki
RAND Corp.

Mildred S. Dresselhaus
Massachusetts Institute of
   Technology

Hugh T. Patrick
Columbia University

John D. Rockefeller IV
United States Senate

David A. Duke
Corning, Inc.

Robert A. Scalapino
University of California, Berkeley

James M. Fallows
The Atlantic

Susan C. Schwab
Motorola, Inc.

Daniel J. Fink
D. J. Fink Associates, Inc.

Ex Officio Members:

Gerald P. Dinneen, Foreign Secretary, National Academy of Engineering

James B. Wyngaarden, Foreign Secretary, National Academy of Sciences
   and Institute of Medicine

iii

WORKING GROUP ON U.S.-JAPAN TECHNOLOGY LINKAGES IN TRANSPORT AIRCRAFT

Daniel J. Fink, *Chairman*
D. J. Fink Associates, Inc.

Lawrence W. Clarkson
The Boeing Co.

Thomas M. Culligan
McDonnell Douglas

Jacques S. Gansler
TASC (The Analytic Sciences Corp.)

John R. Girotto
Collins Commercial Avionics

Jim C. Hoover
Northrop Corp.

Lee Kapor
GE Aircraft Engine Group

Donald H. Lang
Pratt & Whitney

Edward J. Lincoln[*]
The Brookings Institution

Richard J. Samuels
Massachusetts Institute of Technology

Robert M. White
Carnegie Mellon University

[*]Edward Lincoln is currently Special Economic Advisor to the U.S. Ambassador to Japan.

OFFICE OF JAPAN AFFAIRS

Since 1985 the National Academy of Sciences and the National Academy of Engineering have engaged in a series of high-level discussions on advanced technology and the international environment with a counterpart group of Japanese scientists, engineers, and industrialists. One outcome of these discussions was a deepened understanding of the importance of promoting a more balanced two-way flow of people and information between the research and development systems in the two countries. Another result was a broader recognition of the need to address the science and technology policy issues increasingly central to a changing U.S.-Japan relationship. In 1987 the National Research Council, the operating arm of both the National Academy of Sciences and the National Academy of Engineering, authorized first-year funding for a new Office of Japan Affairs (OJA). This newest program element of the Office of International Affairs was formally established in the spring of 1988.

The primary objectives of OJA are to provide a resource to the Academy complex and the broader U.S. science and engineering communities for information on Japanese science and technology, to promote better working relationships between the technical communities in the two countries by developing a process of deepened dialogue on issues of mutual concern, and to address policy issues surrounding a changing U.S.-Japan science and technology relationship.

Staff

Alexander De Angelis, Director[*]
Thomas Arrison, Research Associate
Maki Fife, Program Assistant

---

[*]Alexander De Angelis assumed the position of Director of the Office of Japan Affairs after the departure of Martha Caldwell Harris.

Contents

APPENDIXES

Executive Summary

OVERVIEW

For more than 50 years, U.S. leadership in aircraft manufacturing and aviation has been a major component of our economic strength and national security. Today, that leadership is being challenged as U.S. aircraft primes[1] and a broad range of suppliers face depressed commercial markets, cuts in defense spending, and intense international competition. As markets, capital, and technological capabilities become increasingly global, international strategic alliances and other cross-border linkages have become a familiar feature of this industry. The importance of Japan and Japanese companies for the U.S. aircraft industry—as partners, customers, and competitors—is already substantial and growing rapidly.

It is in this environment of upheaval and opportunity that the National Research Council's Committee to Assess U.S.-Japan Technology Linkages in Transport Aircraft has examined the context, current status, and implications of U.S.-Japan relationships that develop or transfer aircraft technology. Although the European consortium Airbus Industrie is the only existing foreign prime for large commercial transports, this study of U.S.-Japan linkages is timely and appropriate for several reasons. To begin with, Japan's participation in the global aircraft industry is more extensive than is generally recognized, and has been achieved largely through alliances with U.S. industry. Further, the technological capability of Japan's aircraft industry is rising rapidly. If we look beyond the existing competition between primes to an industry constituted of sophisticated parts and subsystem suppliers, Japan's importance becomes more evident. Finally, other countries may seek to emulate Japan's strategies for aircraft industry development in the future. Therefore, a focus on U.S.-Japan relationships carries important implications for how overall competitive challenges will evolve in this industry.

---

[1] In this report, the term "primes," refers to prime contractors.

1

From its assessment, the committee concludes that although continuing and expanded U.S.-Japan cooperation is inevitable and consistent with U.S. interests, a new approach is needed to ensure that this cooperation contributes to reenergizing U.S. leadership in the aircraft industry.

MAJOR FINDINGS

1. *Leadership in global competition will increasingly go to the firms emphasizing high-quality, low-cost manufacturing. This is precisely the area that the Japanese have made their top priority—at the same time that the U.S. aircraft industry is making deep cuts in capital equipment investment.*

A major purpose of the assessment was to reexamine the widely held assumption that Japan's aircraft industry is unlikely to move into the ranks of global leaders. During its study mission to Japan in June 1993, the committee was impressed by the investments in state-of-the-art manufacturing equipment made by the four "heavies" that lead Japan's aircraft industry—Mitsubishi Heavy Industries (MHI), Kawasaki Heavy Industries (KHI), Ishikawajima-Harima Heavy Industries (IHI), and Fuji Heavy Industries (FHI)—as well as by dedicated aircraft suppliers such as Teijin Seiki. Japanese industry has attained world-class capabilities in manufacturing aircraft components such as fuselage panels, thick and complex composite structures, long shafts for aircraft engines, and primary actuation. Advanced technologies—including processes utilizing five-axis machines driven by design data bases—are combined with a meticulous approach to manufacturing practice to achieve high quality, low cost, and reduced cycle time. In addition, the manufacturing excellence achieved by companies such as Toray in carbon fiber and Sharp in flat panel displays has allowed Japanese industry to establish dominant positions in several critical areas of the aircraft supplier chain.

Although some U.S. companies are making the investments necessary to stay on the cutting edge of manufacturing, many are not, largely as a result of ongoing cuts in military aircraft procurement and the current commercial market slump. This is a critical issue because the forces shaping competition over the next decade—growing but price-sensitive markets, industry restructuring, and fewer brand-new aircraft and engine programs than in the past—mean that both U.S. primes and suppliers will be continually pressured to deliver more value at lower cost.

2. *Japan's aircraft R&D and defense production systems actively foster an integrated and flexible dual-use technology and manufacturing base. In contrast, the commercial benefits of U.S. military aircraft R&D and procurement have declined over time—largely as a result of policies that implicitly discourage military-commercial synergies.*

Although the amount of public resources expended on the Japanese aircraft industry is relatively small, Japan's government-sponsored domestic cooperative programs are more strongly oriented to technology sharing among

Japanese companies and commercialization of aircraft technologies than those supported by the U.S. government. For example, the Key Technology Center project on aluminum-lithium alloys launched in 1989 provides investment funding to aluminum manufacturers and fabricators for research likely to have important applications in the aircraft industry. In the United States, the National Aeronautics and Space Administration (NASA) aeronautics program has produced many advances that have enhanced the competitiveness of U.S. firms in the past, but there has been no comprehensive effort directed toward technology commercialization and product application technology.

Japanese and U.S. policies on the defense side provide contrasts as well. For example, Japan's defense R&D spending has a strong dual-use orientation, while U.S. military aircraft development increasingly emphasizes unique capabilities that enhance combat performance—such as stealth and high maneuverability—but have few direct commercial applications. In addition, the production facilities of Japanese companies often manufacture components for both military and commercial aircraft side by side or even on the same equipment, whereas U.S. companies such as McDonnell Douglas and Boeing have found it prudent to separate similar military and commercial manufacturing because of procurement regulations and unique military specifications. Although the current administration is initiating efforts to change such regulations, the existing system inherently separates (rather than integrates) military and civilian R&D and production.

3. *Japan uses international partnerships strategically to foster technology acquisition. Japan's policy and business environment allows industry to gain maximum leverage from international alliances, resulting in a gradual upgrading of independent technological capabilities and a diffusion of these skills throughout the manufacturing network of primes and suppliers.*

The Japanese aircraft industry does not carry out full independent integration of commercial airframes, engines, or avionics, but it has achieved increasing independence and growing technological strength by promoting international linkages. The Japanese government supported the JFR-710 project in the 1970s, laying the foundation for Japanese industry's participation in the multinational V2500 engine program. More recently, the Ministry of International Trade and Industry (MITI) launched the HYPR program in 1989, designed from the start as an international collaborative effort in advanced, supersonic engine technologies. The Japanese government has also supported and coordinated Japanese participation in Boeing's aircraft programs. MHI, KHI, and FHI have increased the extent and technical sophistication of their relationship with Boeing over time.

While the Japanese policy process for international partnerships is oriented toward "behind the scenes" advance government-industry coordination, the U.S. policy process is more ad hoc and uncoordinated. This contrast is particularly important now since both the U.S. and Japanese aircraft industries are being forced by changes in the global environment to make significant adjustments. International linkages are very much a focus of current Japanese planning, as a series of new studies, working groups, and international missions have been organized in past months to consider critical decisions relevant to the future of the industry. MITI and industry are jointly developing new approaches to strengthen Japan's aircraft industry for the twenty-first century. There is no such effort under way in the United States.

4. *The U.S. aircraft industry has gained significant benefits from its relationships with Japan, including sales and licensing income, world-class components, and financial leverage for costly new programs. Yet cooperation entails risks and raises concerns as well, particularly the long-term impacts of technology transfer from the United States to Japan and the effects of linkages on the U.S. supplier base.*

Although the predominant flow of technology in U.S.-Japan aircraft alliances has been from the United States to Japan, the U.S. Department of Defense (DOD) and U.S. companies involved in military and commercial linkages have structured programs with the aim of protecting critical technologies. Still, the impacts of the most recent technology transfers are unclear. Japanese industry is not competing today at the prime integrator level, but it already possesses or could acquire the capabilities needed to do so. In addition, Japanese companies are displacing U.S. suppliers in areas such as fuselage structures, and they dominate several critical component technologies. While Japan does not have offset requirements or other formal market barriers, U.S. manufacturers selling to Japan feel informal pressure to source there in order to enhance access to Japanese airlines, and some have found it difficult to participate in the Japanese market without a joint venture with a Japanese company.

Rather than retreat into a "protectionist" or defensive stance, the United States should pursue a proactive approach to building effective U.S.-Japan relationships that involve a more balanced flow of aircraft technologies between the two countries. Further, there is a need to promote more effective working relationships between U.S. companies and between industry and government to ensure the retention of an innovative, full-spectrum aerospace capability in the United States.

IMPERATIVES FOR THE FUTURE

Although this study focuses on Japan, it is clear that U.S. leadership has been and will continue to be challenged by other industrialized countries that view aviation as fundamental to their economic growth. The committee developed future scenarios for the course of the global aircraft industry and U.S.-Japan alliances over the next decade and beyond. Several scenarios that contemplate declining U.S. market share are plausible and could come about if current trends continue. *The most desirable scenario—a resurgent, globally competitive U.S. aircraft industry—will not be realized unless U.S. companies and government work together to bring about a significant change in course.*

Leadership in aircraft design and manufacturing—including a full-spectrum supply chain—remains a vital U.S. national interest. As a result of its assessment, the committee concludes that in order for the United States to maintain its leadership in this critically important industry, government-industry partnering in the development and implementation of a long-term strategy is essential. While the major responsibility lies with the U.S. aircraft industry itself, government must do more to create a favorable overall environment. Currently, neither a coherent policy nor the needed institutional mechanisms exist.

POLICY RECOMMENDATIONS

In response to the need for a comprehensive and proactive U.S. approach, the committee has developed a five-part strategy outlining the critical imperatives for U.S. industry and government attention, along with specific action items.[2] The five elements are

1. maintaining U.S. technological leadership,
2. revitalizing U.S. manufacturing capabilities,
3. encouraging mutually beneficial interaction with Japan,
4. ensuring a level playing field for international competition, and
5. developing a shared U.S. vision.

Maintaining U.S. Technological Leadership

The current massive restructuring on both the military and the commercial sides of the aircraft business makes it critical that U.S. technological leadership be maintained. NASA must continue to play a key role in aeronautics. Its currently proposed 35 percent increase in aeronautics funding should continue for three more years. NASA's traditional role in basic research should be expanded into product-applicable technologies in subsonic aeronautics and propulsion systems, with the primary objective of reducing the investment and operating costs of future aircraft systems. To ensure increasing commercial application of these technologies, NASA should increase significantly the funding share contracted to industry. Also, the U.S. Department of Defense should maintain its aircraft R&D budget for enabling technologies at current levels despite overall cuts in the defense budget.

U.S. industry must continue to invest its own resources in new technology development. In order to facilitate this investment, the R&D tax credit should

---

[2]An abridged set of recommendations is presented here. The complete list is contained in Chapter 5.

be made permanent, and incentives should be developed to avoid penalizing companies that reorient their R&D from defense-unique to dual-use or commercial areas.

Revitalizing U.S. Manufacturing Capabilities

U.S. primes and suppliers will have to improve manufacturing performance continually in terms of cost, quality, and delivery to remain competitive—especially in view of the large investments in state-of-the-art equipment being made by the Japanese aircraft industry. To this end, a well-structured investment tax incentive designed to encourage productivity-enhancing investments should seriously be studied, both for its practicality and effectiveness, and compared to the incentives provided to industry in Japan and Europe.

Department of Defense reform of its procurement system is the key to promoting greater civil-military integration, especially in the area of reducing barriers to common R&D and manufacturing facilities for military and civilian aircraft production. Reform of the system should include more extensive use of commercial item descriptions, a greater emphasis on low cost and high quality, and revisions in accounting standards. R&D funding by DOD should place a high priority on manufacturing and design processes, and give priority to cooperation between primes and suppliers in U.S. government RFPs (Request for Proposals). DOD should also consider carrying aircraft and subsystem prototypes forward to limited production in order to demonstrate low-cost "manufacturability" as well as performance.

Encouraging Mutually Beneficial Interaction with Japan

The environment surrounding U.S.-Japan linkages has evolved significantly, demanding a new approach to ensure that the benefits of cooperation are maximized and the risks are managed. As part of its activities to promote greater reciprocity in the transfer of aircraft technology between the United States and international partners—including Japan—a private sector effort should be launched to identify critical technologies, establish guidelines covering the transfer of commercial aerospace technology, and periodically assess international technology transfers. The Department of Commerce should also consider leading a new initiative to collect and disseminate technical and business information from global sources to the U.S. aircraft industry, including expanded technology benchmarking.

Ensuring a Level Playing Field for International Competition

In light of heightened international competition in all segments of the aircraft industry and the context of heavy government involvement, U.S. trade policy should aggressively promote fair global market competition. The U.S. government should work closely with industry and other governments to achieve multilateral rules that govern and reduce subsidies in this industry. The recently increased Export-Import Bank guarantee and loan activity should also be maintained.

Developing a Shared U.S. Vision

The four previous strategic elements and their associated recommendations are important ingredients for a reenergized U.S. aviation industry with enduring global leadership. What continues to be missing is an institutional mechanism that is committed to the further development and refining of a U.S. aviation strategy, that can understand and include the views of all the necessary players, and that has the visibility and persuasive powers to champion implementation. There is no present government agency that has singular responsibility for the aviation infrastructure. There is no U.S. equivalent of MITI, nor should there be. In any event, it is the private sector that is ultimately responsible for the success or failure of any aviation strategy.

The committee explored several alternative mechanisms for developing a shared vision, such as organizational changes in government, utilization of an existing advisory panel, or tasking one or more industry associations, and found all of these approaches wanting. Accordingly, its final recommendation is the establishment of a National Aviation Advisory Committee (NAAC), composed of knowledgeable leaders from industry, academia, and elsewhere, reporting to the National Economic Council or an interagency group of senior officials. The committee believes that the stature of its membership coupled with its strategic reporting level would help ensure knowledgeable input from the private sector to government councils, as well as a higher likelihood of a coordinated approach for an industry where the United States needs to retain world leadership. The committee recognizes that such a recommendation might be viewed as self-serving for a particular industry, and is aware of problems and mixed effectiveness of similar high-level advisory committees for other sectors. Nevertheless, during this period of restructuring following the ending of the Cold War and with increasing frictions in high-technology competition between the United States, Japan, and Europe, the committee believes that maintaining U.S. leadership in the aviation industry requires a careful assessment and a focused strategy from both U.S. industry and government. This report outlines some of the specific tasks that need to be accomplished. The NAAC as outlined here could perform these tasks as well as address the overriding need for a shared U.S. strategic vision for a continually reenergized leadership position in aviation.

1

Introduction

The strong position that the U.S. aviation industry holds in the world today represents one of America's great industrial success stories. The U.S. aerospace industry, a major exporter, supplies more than half of the world market and ranks sixth among U.S. industries in total sales. (See Appendix A for an analysis of the importance of the U.S. aircraft industry.) Many of the competencies built in this R&D-intensive industry diffuse to other industries and contribute to the overall economy.

The industry is in a real sense a major national asset. The U.S. leadership position in aircraft is the result of a continuous stream of investments in new technologies across a broad spectrum. Substantial support has come from government-funded projects that have spun off commercial applications—the J52 engine formed the core for the Pratt & Whitney JT8D engine on the DC-9; the core of the GE F110 engine for the F-16 was used as a basis for development of the CFM56 engine for the 737, A320, A321, and A340. Commercial aircraft are tested in National Aeronautics and Space Administration (NASA) wind tunnels, and NASA's work in areas such as computational fluid dynamics helped Boeing locate the nacelles on the wings of the 737, 757, and 767 to minimize drag.[1] At the same time, technology employed in commercial transports is often used in military programs, and commercial aircraft produc-

---

[1] See U.S. Congress, Office of Technology Assessment (OTA), *Competing Economies—America, Europe and the Pacific Rim* (Washington, D.C.: U.S. Government Printing Office, 1991), p. 347.

tion increasingly contributes to maintaining the supplier and work skill base, and produces cost economies for companies that manufacture both civilian and military aircraft.

The critical questions for this study are whether the United States can maintain its lead in the future, and the likely impacts of U.S.-Japan technology transfer and engineering relationships, broadly defined as technology linkages.[2] Japan's aircraft industry has generally been assumed unlikely to move into the ranks of the global leaders. A major purpose of this study, carried out by a committee of individuals with considerable experience in the industry and knowledge of Japan, was to reexamine that assumption and to look ahead to the future. The third in a series of studies on technology linkages organized by the National Research Council's Committee on Japan, this study, which included a committee study trip to Japan, was carried out during 1993 with support from the Defense and Commerce Departments and from the Japan-United States Friendship Commission.

A number of important contextual changes suggest that the future will be different from the past. Global competition is intensifying. Airbus rapidly increased its sales in Japan in 1991 and 1992, overtaking McDonnell Douglas.[3] Industry experts predict that Asia will play a major role in global demand in the 1990s and the first decade of the next century.[4] Over the next decade, or until new technology developments permit the introduction of supersonic and hypersonic transport aircraft, the committee believes that *leadership in global competition will increasingly go to the firms emphasizing high-quality, low-cost manufacturing. This is precisely the area that the Japanese have made their top priority.*

A major transformation is occurring in the industry as defense spending declines with the end of the Cold War. In the past, U.S. defense procurement drove R&D and capital spending in important segments of the industry and aircraft-related technologies. Today, the U.S. aircraft industry is struggling to adjust to these historic changes in a difficult context—a downturn in demand for commercial transports during the past few years. In Japan, where the U.S.-Japan alliance has formed the cornerstone of Japan's defense policy, declines in military procurement are also beginning to force hard choices.[5] President

---

[2]Technology linkages include company-to-company activities (sales and maintenance agreements, licensed production, joint production or development, equity arrangements), as well as relationships involving governments and universities. See National Research Council, *U.S.-Japan Technology Linkages in Biotechnology* and *U.S.-Japan Strategic Alliances in the Semiconductor Industry* (Washington, D.C.: National Academy Press, 1992), for a detailed discussion of the term and approaches to analysis.

[3]These data were provided by GE Aircraft Engines. Airbus has sold to Japan Air Systems, a relatively new domestic airline. Japan Airlines (JAL) and All Nippon Airways (ANA) generally continue to purchase Boeing airplanes.

[4]Boeing Commercial Airplane Group, *1993 Current Market Outlook*, March 1993, p. 3.5.

[5]Japan Defense Agency (JDA) officials emphasized this point in discussions with the committee during their trip to Japan in June 1993. For example, AWACS purchases will comprise 30 percent of Japan's aircraft procurement budget in the next few years. In August of 1993 it was reported that the

Clinton has made it clear that the United States will maintain its military presence in Asia, at the same time working with Japan to build new multilateral approaches to security in the region. However, this is a time of political change in Japan accompanied by a reexamination of fundamental principles. This climate of change and uncertainty is the larger context within which the U.S. transport aircraft industry must compete and cooperate with Japan.[6]

Although this study is primarily concerned with Japan, it is clear that U.S. leadership has been and will continue to be challenged by other industrialized countries that view aviation as fundamental to their economic growth. This report contains frequent references and comparisons to Europe and other parts of the world. The central issues dealt with in this study are generic ones—what are the benefits and what are the risks associated with expanding technological linkages? The committee begins with the premise, well substantiated by previous National Research Council (NRC) studies, that international technological linkages are a fact of life. In the aircraft industry the primary U.S. participants are private companies who seek investment partners, entry to markets, reliable suppliers with world-class manufacturing, and cooperators in new technology development. Japan is the world's second largest country market for aircraft, most of them purchased from U.S. firms.[7]

Although the committee did not study linkages with other countries in the same depth as those with Japan, overall the linkages and alliances that the U.S. aircraft industry undertakes with Japan are more significant—in both business and technological terms—than linkages with any other single country.[8] On the commercial side, for example, the links that U.S. airframe manufacturers have with industries in China and Italy do not involve the extensive design collaboration that exists in Boeing's Japanese alliances. On the military side, Japan is still the only ally that has been allowed to produce the McDonnell Douglas F-15 under license, and most experts agree that the extensive interaction and technology flow contemplated in the FS-X program go far beyond what has been attempted in collaborative programs with other countries.

A major motivation for U.S. linkages with Japanese firms—market leverage—is analogous to the motivation driving military offset deals concluded by

---

Japanese Diet approved a plan drafted by the Japanese Ministry of Finance that will limit increase in JDA's budget for FY 1994 to about 1.9 percent (about $818 million) over the FY 1993 budget of $42 billion. See Barbara Wanner, "Defense White Paper Stresses Regional Threats," *JEI Report*, No. 31B, August 20, 1993, p. 3.

[6]In June of 1993, the *Nihon Keizai Shimbun* reported that Kawasaki Heavy Industries had agreed to provide British Aerospace with advanced production control techniques for application to a missile production facility. See "Kawaju no Kanri Gijustu Donyu" (Introducing KHI's Management Technology), *Nihon Keizai Shimbun*, June 8, 1993, p. 1.

[7]The cumulative total of deliveries of jet airplanes to Japan through 1992 was $32.6 billion, with Boeing providing the bulk of them (data provided by Boeing).

[8]The significance of U.S.-Japan linkages varies across segments in relative terms. For example, although U.S.-Japan alliances are extensive and important in aircraft engines, the CFM International joint venture between General Electric and Snecma of France is clearly the most significant international link by U.S. industry in this segment of the industry.

U.S. aircraft companies, and in some cases joint ventures involving companies from other countries. However, the available evidence indicates that no other country has achieved the level of success that Japan has thus far in leveraging international alliances to build and sustain a domestic aircraft industry.[9] This is because Japan's significance as a market and strategic partner has given it more leverage, and because the Japanese aircraft industry—working closely with the Japanese government—has taken better advantage of the opportunities afforded by alliances. From the Japanese perspective, a significant share of overall aircraft industry sales is derived from projects involving a U.S. linkage.[10]

Perhaps the main concern is that these linkages will, however, result in the building of strong commercial competitors by expanded transfer of U.S. technology abroad. Although normally framed in terms of the potential emergence of new airframe integrators, the downside risks affect even more directly the U.S. suppliers of subsystems and components, some subsegments of which are already losing market share to foreign firms. The U.S. aircraft industry, broadly defined to include the networks of related technical expertise and manufacturing capabilities that link the primary manufacturers and the suppliers, is a major national asset. The focus of this report is Japan—as a partner in both cooperation and competition—but the questions are generic, and it is hoped that the answers will contribute to building effective national policy, public and private, for the twenty-first century.

The chapters that follow provide a summary of the committee's analysis and recommendations. Chapter 2 provides a brief introduction to the historical evolution of Japan's aircraft industry and the overall policy context in Japan and the United States. Chapter 3 analyzes U.S.-Japan technology linkages in transport aircraft and draws conclusions about impacts on the United States. Chapter 4 outlines alternative scenarios for the future. Chapters 5 outlines policy issues and recommendations. Readers are encouraged to refer to the appendixes of this report for detailed information and assessment.

---

[9]The members of Airbus Industrie have, of course, taken a very different path. They have leveraged their existing domestic capabilities in pursuing global market share through a multinational alliance.

[10]Between 1987 and 1991, Japan's aerospace exports to the United States more than doubled. See Aerospace Industries Association, *Aerospace Facts and Figures 1992-1993* (Washington, D.C.: AIA, 1992), p. 122.

2

Background and Policy Context

HISTORICAL BACKGROUND

Aircraft development has always been a high-risk, demanding business. Historically, new product development costs have often exceeded the market value of the company making the investment. Currently, development costs for a major new program (such as the Boeing 777) may exceed $5 billion.[1] The economics of the aircraft industry push toward international linkages formed to share risk.

Powerful counterforces, however, explain the desire for an indigenous aircraft industry in many nations where entry into the airframe integration segment of the industry is economically irrational from the perspective of any individual company. Throughout the history of the Japanese aircraft industry there has been an interplay between the push for indigenously developed technologies by an independent Japanese industry, and the need to form technology linkages, given the realities of the global marketplace and the need to access technology from abroad.

The four Japanese "heavies"—Mitsubishi Heavy Industries (MHI), Kawasaki Heavy Industries (KHI), Fuji Heavy Industries (FHI), and Ishikawajima-Harima Heavy Industries (IHI)—that dominate Japan's aircraft industry today have all been involved in aircraft production since the early part of the twenti-

---

[1]See "Betting on the 21st Century Jet," *Fortune*, April 20, 1992, p. 102.

13

TABLE 2-1 Selected Japanese Aerospace Manufacturers
(Estimated FY 1992 million dollars, ¥110 per dollar)

Company	Sales	Aerospace Sales (% of total)		Corporate R&D
MHI	22,545	3,382	(15%)	1,064
KHI	8,636	2,245	(26%)	209
IHI	7,272	1,236	(17%)	340
FHI	7,909	474	(6%)	227
Four Heavies, total	46,362	7,337	(16%)	1,840
Toray	5,409	?		291
Shimadzu	1,569	439	(28%)	118
Teijin Seiki	623	224	(36%)	19
JAE	564	118	(21%)	26
Nippi	300	288	(96%)	3.6
Selected suppliers total	8,465	?		458

NOTE: Companies do not provide breakout figures for aerospace or aircraft-related R&D.

SOURCE: Compiled by Office of Japan Affairs from data appearing in Toyo Keizai, *Japan Company Handbook—First Section* (Tokyo: Toyo Keizai, 1993).

eth century. Today they manufacture structural parts of aircraft and act as risk-sharing partners for large aircraft and engine development projects led in most cases by foreign-based firms (see Table 2-1). In addition to the four heavy industry companies that lead Japanese participation in commercial programs and act as prime contractors for major weapons systems purchased by the Japan Defense Agency (JDA), the Japanese aircraft industry consists of many subcontractors as well as many companies that have developed competitive capabilities in the manufacture of various aircraft components. In a number of cases, these companies, such as Toray, are applying technologies developed for another market segment. The United States has become increasingly dependent on Japanese suppliers for some types of components, such as flat panel displays. A distinguishing feature of the Japanese industry is its strength in components supply.

Japan's aircraft industry is also distinguished by its reliance on military production (see Table 2-2).[2] *In 1991, defense production accounted for almost 75 percent of Japan's total aircraft industrial output.*[3] At the same time, it is

[2]For FY 1992, JDA aircraft procurement was $2.46 billion (¥270 billion at ¥110/$1) versus U.S. Department of Defense aircraft procurement of $23.95 billion (estimated). See Boeicho (JDA), *Heisei Yonendo Boeihakusho* (Tokyo: Okurasho Insatsu Kyoku, 1992), p. 302; and Aerospace Industries Association (AIA), *Aerospace Facts and Figures 1992-1993* (Washington, D.C.: AIA, 1992), p. 22.

[3]See Nihon Kokuchukogyokai (Society of Japanese Aerospace Companies), *Heisei Yonendohan Kokuchunenkan* (Aerospace Industry Yearbook 1992 Edition), (Tokyo: Koku Nyusu, 1992), p. 433.

TABLE 2-2 U.S. and Japanese Aircraft Industries—1991 Sales and Trade Comparison (million dollars, ¥110 per dollar)

	United States	Japan
Total aircraft sales	68,593	7,735
Sales to domestic government[a] (% of total)	21,703 (32%)	5,926 (77%)
Aircraft imports	12,626	5,127
Aircraft exports	42,412	841
Aircraft trade balance	29,786	-4,286

NOTE: [a]For both countries, nearly all domestic government sales are military.

SOURCE: Nihon Kokuchukogyokai (Society of Japanese Aerospace Companies), *Heisei Yonendohan Kokuchukogyo Nenkan* (Aerospace Industry Yearbook 1992 Edition), (Tokyo: Koku Nyusu, 1992), pp. 433-437; and Aerospace Industries Association, *Aerospace Facts and Figures 1992-1993* (Washington, D.C.: AIA, 1992), pp. 28, 126.

important to note that the same major Japanese companies that produce aircraft engage in widely diversified production of ships, nonaircraft military vehicles and engines, and missiles, as well as nonaerospace production in areas such as motorcycles, electronic devices, and textiles. Industrial diversity is a hallmark of the large companies, a characteristic that enhances synergies between military and civilian production that are unusual in the United States and explicitly encouraged by Japanese government policies.

Japan's pre-World War II industry was promoted for national purposes, and organized to acquire needed foreign technologies through licensing and other linkages to foreign companies while at the same time building domestic Japanese capabilities through government-directed procurement, R&D, and planning.[4] Japan's success with the Mitsubishi A6M5, popularly known as the Zero fighter, demonstrated the high level of domestic capabilities spawned by the "independent aircraft policy" of the 1930s.

Japan's aircraft industry, which had been one of the largest and most technologically advanced in the world during World War II, was initially prohibited production by the American occupation after the war. In the early postwar period, the industry was formally dismantled, and some of its accumulated technical and human expertise flowed to other Japanese industries such as automo-

[4]For a detailed analysis, see Richard J. Samuels, chapter 4, "The Japanese Imperial Aircraft Industry," *Rich Nation, Strong Army: National Security, Ideology, and the Transformation of Japan* (Ithaca, N.Y.: Cornell University Press, forthcoming 1994).

biles.[5] The revival and expansion of the industry were made possible under the U.S.-Japan security treaty and were given their first stimulus during the Korean War when the U.S. military contracted with Japanese firms for significant maintenance and repair work.

The major mechanism for expansion of the military aircraft industry was licensed production in Japan of U.S.-designed aircraft, despite the considerably higher costs of production versus purchase of U.S.-made aircraft. Over time, Japanese firms progressed in defense production from assembly of U.S.-fabricated "kits" to production of more components of greater sophistication. The FS-X program represents a new stage of joint development, with the Japanese firm MHI acting as the prime contractor and Japanese firms taking on a much larger role in design from the outset. Independent Japanese programs have centered on trainers and day fighters, rather than the highest-technology military aircraft (see Table 2-3). Over the past 40 years or so, Japan has pursued an incremental approach to building its industry by maximizing and expanding its participation through linkages primarily with U.S. firms and consistent with U.S. government policy encouraging military cooperation.

The Aircraft Promotion Law of 1958 established the policy framework for promotion of commercial aircraft production. Although Japan did make one attempt (the YS-11) to develop a commercial aircraft, the 64-seat, twin-engine turboprop was a failure in the market.[6] Since that time, all major commercial transport aircraft programs in which Japan has participated have involved technology linkages with foreign firms. Japanese firms progressed from work as subcontractors on Boeing's 747, 727, and 737 models and on McDonnell Douglas's DC-9 and DC-10 during the late 1960s and early 1970s to "risk-sharing subcontractors" involved in the development and production of the 767 in the late 1970s. These linkages are explored in more detail in the next chapter.

U.S. AND JAPANESE POLICIES

Japanese and U.S. government policies toward the aircraft industry provide striking contrasts. The contextual changes mentioned above are forcing adjustments in both countries, and in many ways the 1990s are a watershed period. Critical choices made today will have significant impacts for many years to come.

As a basis for comparison, a number of policy vehicles are examined briefly, along with the overall process of decision making. The policy instruments include direct financial assistance, support for civilian R&D, military and civilian procurement synergies, and other forms of government action to

---

[5]Richard Samuels points out that despite the formal ban, 40 percent of the facilities were maintained and 80 percent of the engineers stayed on at IHI and Nakajima. Ibid., chapter 7.

[6]Only 182 planes were sold and the government forgave a large debt. MHI and FHI made several attempts to enter the smaller business-class aircraft segment without success.

TABLE 2-3 Japan's Postwar Military Aircraft Programs (Excluding Helicopters)

Program	Manufacturer/Partner, Linkage	Period	Number Produced
Independent Japanese Programs			
T-1A/B	Fuji Heavy Industries (FHI)	1955-1962	60
PS-1	ShinMaywa Industries	1965-1978	23
US-1	ShinMaywa Industries	1973-present	14+
T-2	Mitsubishi Heavy Industries (MHI)	1970-1987	97
F-1	Mitsubishi Heavy Industries	1974-1976	77
C-1	Kawasaki Heavy Industries (KHI)	1971-1981	29
T-4	Kawasaki Heavy Industries	1985-present	76+
Programs with American Participation			
F-86F	MHI-North American, licensed production	1955-1960	300
T-33	KHI-Lockheed, licensed production	1954-1959	210
F-104J	MHI-Lockheed, licensed production	1960-1966	210 (20 FMS)
F-4EJ	MHI-McDonnell, licensed production	1968-1980	138 (2 FMS)
F-15J/DJ	MHI-McDonnell, licensed production	1977-present	250+ (14 FMS)
P-3C	KHI-Lockheed, licensed production	1978-present	75+
FS-X	MHI-Lockheed, codevelopment	1987-present	?

NOTE: "FMS" refers to the U.S. foreign military sales program—these aircraft were direct sales undertaken in addition to those produced under license.

SOURCE: Compiled from various sources by the National Research Council Working Group on U.S.-Japan Technology Linkages in Transport Aircraft.

organize the industry, set overall national goals, and develop the aviation infrastructure.

In Japan, direct and indirect financial assistance has been an important policy instrument for government support of the commercial aircraft industry. The second Aircraft Promotion Law of 1958 set the policy framework for promoting the civilian aircraft industry. One concrete manifestation was the organization by the Ministry of International Trade and Industry (MITI) of the Nippon Aircraft Manufacturing Company, a consortium of MHI, KHI, FHI, ShinMaywa Industries, Showa Aircraft, and Japan Aircraft, to build the YS-11 in which the government held half the equity. MITI provided more than half of the development costs and even guaranteed coverage of losses that the companies incurred in the production phase. Another example was the formation in 1971 of Japan AeroEngines, a consortium of IHI, MHI, and KHI, to develop a high-bypass engine. Once again, MITI covered half the development costs with success-conditional loans.[7] Beginning in the early 1970s, MITI provided success-conditional loans for Japanese partnerships with Boeing as risk-sharing subcontractors in a development program that ultimately became the Boeing

---

[7]Success-conditional loans are repaid as the borrower earns revenue on the targeted project.

767.[8] In 1986, the Japanese government supported the initiation of Japanese partnership with Boeing in the 7J7 project (later put on hold) to develop a narrow-bodied civil transport. In FY 1993, the government of Japan reportedly provided 2 billion yen ($16 million) for the 777 project,[9] as well as loans from the Japan Development Bank (JDB) and the Export-Import Bank for development and for aircraft imports.[10]

The International Aircraft Development Fund (IADF), since its establishment in 1986, has been a major vehicle for government support of Japanese participation in new international commercial aircraft programs. Establishment of the IADF reflected MITI's decision in the 1980s to foster international collaboration as the major mechanism for strengthening Japan's domestic aircraft industry. The IADF, supported by corporate member contributions and indirect government aid,[11] distributes interest-free loans that must be repaid out of revenue from the project.[12] Although the Japanese financial contributions to a large project such as the 777 make up only a portion of the total, it is a significant portion. The Japanese are risk-sharing partners developing 20 percent of the airframe for the 777 project, which may cost as much as $5 billion; in addition, support from the government (in the form of loans from the JDB and indirectly through the IADF) may well total $300 million to $400 million annually for the project, not to mention JDB and Export-Import Bank loans for aircraft imports that provide revenues to Boeing. Direct financial support brings benefits to foreign as well as Japanese firms (see Table 2-4).

Direct financial assistance to the commercial aircraft industry has not been a major U.S. policy instrument. During the postwar period, such assistance has been extended on three occasions—$1 billion for development of the supersonic transport in the 1960s, and loan guarantees (never actually called upon) to two struggling aircraft producers. The Office of Technology Assessment (OTA) observes that these examples "pale in comparison" to direct financial assistance by other governments and that the interventions "were ad hoc, not a part of a coherent strategy to support the commercial aircraft industry."[13]

Reflecting differences in government policies and corporate practices, the Japanese aircraft industry made comparatively large investments in capital spending. In 1990 the U.S. aerospace industry invested $3.4 billion (2.7 percent of sales) in capital spending, while the Japanese aerospace industry spent $1

[8]Loans from MITI totaled 14.7 billion yen for the development phase. According to MITI officials, the loans were more than 90 percent repaid by the summer of 1993.

[9]*Wing Newsletter*, January 13, 1993, p. 7.

[10]JDB loans in the amount of almost $1 billion were allocated for the V2500, 777, and 7J7 projects.

[11]In FY 1992, 4.3 billion yen (about $40 million) was provided through MITI's budget for the V2500, 777, and YXX programs. See Nihon Kokuchukogyokai, op. cit., p. 426.

[12]See Samuels, op. cit., chapter 7, for a more detailed analysis of Japanese government support for the Japanese aircraft industry.

[13]U.S. Congress, Office of Technology Assessment, op. cit., p. 348.

TABLE 2-4 Fiscal 1993 Japanese Government Aircraft Industry Support
(million dollars, ¥110 per dollar)

R&D and Program Support	
MITI total	92.5
V2500	19.3
777	18.2
YXX	5.3
HYPR	36.8
SST market studies	1
Advanced heat-resistant materials	16
Small airplane studies	1
Small engine research	0.1
Test facilities	0.1
(program support includes $5.5 million from non-MITI sources)	
Japan Development Bank Loans	
V2500, 777 and YXX	1,091 (1)
Science and Technology Agency	
National Aerospace Laboratory	66 (2)
Japan Defense Agency	1,091 (3)
Other Support	
Japan Development Bank and Export-Import Bank	1,597
loans for aircraft imports	

NOTES: (1) The JDB figure is the total available—it is possible that not all of this will be lent. (2) According to the National Aerospace Laboratory (NAL), NAL budget reporting significantly over-states aeronautics funding because most personnel and overhead costs for both aeronautics and space research are reported under aeronautics. NAL aeronautics R&D funding minus overhead and salaries was about $3 million in 1993. (3) The figure given here is the budget for the Japan Defense Agency's Technical Research and Development Institute (TRDI). In 1992, over half of TRDI's budget went toward research contracted to industry in connection with the FS-X codevelopment program.

SOURCE: *The Wing Newsletter*, January 13, 1993, pp. 7-8; Communication from National Aerospace Laboratory, July 1993; Science and Technology Agency, *Indicators of Science and Technology 1993* (Tokyo: Okurasho Insatsukyoku, 1993), p. 125; and Japan Defense Agency, November 1993.

billion (10 percent of sales).[14] Capital equipment spending will have long-term payoffs in improved production. Furthermore, high capital expenditure encourages important forms of technological change that are not captured in the R&D figures.

Large capital equipment purchases enable the Japanese firms to move quickly in adopting new, advanced production methods. The manufacturers work closely with the equipment suppliers in this process. As discussed later in this report, this focus on technical change in the manufacturing process is con-

---

[14]See AIA, op. cit., p. 160; and Nihon Kokuchukogyokai, op. cit., pp. 430, 437.

19

sistent with the Japanese emphasis on cost and quality in production (rather than on overall product design) (see Tables 2-5 and 2-6).

The Japanese government promotes diffusion of technology through cooperative civilian R&D projects. During the 1980s the government of Japan launched a number of R&D consortia designed to develop new technologies needed in the aircraft industry, particularly engines. The advanced turboprop engine project, for example, was supported as a Key Technology Center project beginning in 1986. The Frontier Aircraft Basic Research Center Company was established to carry out the work with 70 percent equity participation by the Key Technology Center (under MITI and the Ministry of Posts and Telecommunications) and the remaining equity provided by the 34 participating firms,

TABLE 2-5 1991 Capital Spending (million dollars, ¥110 per dollar)

	Total	Percentage of Sales
U.S. aerospace industry	4,040	2.9
Japanese aerospace industry	852	8.2

NOTE: U.S. figure for SIC codes 372 and 376. Japanese figures represent the results of a survey of 24 companies. Both sets of figures may, therefore, undercount total aircraft-related capital expenditure by not including a number of supplier firms.

SOURCE: Aerospace Industries Association, *Aerospace Facts and Figures 1992-1993* (Washington, D.C.: AIA, 1992) p. 160; and Nihon Kokuchukogyokai (Society of Japanese Aerospace Companies), *Heisei Yonendohan Kokuchukogyo Nenkan* (Aerospace Industry Yearbook 1992 Edition), (Tokyo: Koku Nyusu, 1992), pp. 430, 437.

TABLE 2-6 New Plant and Equipment Expenditures by U.S. Business (percentage change from preceding year in current dollars)

	Actual 1991	Actual 1992	Planned 1993 (July-August 1993 survey)
All industries	-0.8	4.6	7.1
Manufacturing	-5.1	-4.8	3.4
Aircraft	0.8	7.6	-22.1

SOURCE: U.S. Department of Commerce, Bureau of the Census.

which included auto and machinery makers and materials fabricators.[15] The project, which ended in 1993, was carried out through distributed research and sharing of results. The project paid for new testing equipment eventually sold to the participants on depreciated terms at the end of the project. Another Key Technology Center project beginning in 1989 focused on fabrication and design technologies for aluminum-lithium alloys. Although aircraft manufacturers are not shareholders, the project provides investment funding to the aluminum manufacturers and fabricators for research likely to have important applications in the aircraft industry. Projects such as these promote the diffusion of know-how not only throughout the aircraft industry, but also through related industries, and divide the research work in ways that create niches of unique expertise for various corporate participants.

International partnerships are used strategically to foster technology acquisition. Japanese government agencies have sponsored two R&D consortia in the engine field. The first, the JFR-710 project, supported by the National Aerospace Laboratory as an experimental development project in the 1970s, provided the foundation for Japanese participation in the V2500 project.[16] More recently, MITI launched the HYPR program in 1989, designed from the start as an international collaborative effort in supersonic engine technologies. Scheduled to continue until 1996, the project is funded by MITI at a level of about $37 million in FY 1993 and administered through MITI's Agency for Industrial Science and Technology and the New Energy and Industrial Technology Development Organization. The aim is research and scale demonstration of a Mach 5, methane-fueled, combined-cycle engine. The major Japanese companies (IHI, MHI and KHI) participate, along with foreign firms, which make up a total of 25 percent participation.[17] The Japanese firms are the lead companies, teaming with foreign firms for various aspects of the development project (see Figure 2-1). The HYPR project is important as Japan's first attempt to organize and lead an international collaborative effort to develop advanced aviation technology. The project is also important because the Japanese government eventually revised its legislation on intellectual property rights, allowing foreign firms ownership in response to jointly organized representation from the foreign firms.[18]

The U.S. government funds R&D for civil applications through the National Aeronautics and Space Administration's (NASA) aeronautics program (see Tables 2-7 and 2-8). Although research supported by NASA has produced many advances, a recent National Research Council (NRC) report concludes that "the attention paid to civil aeronautics in the NASA budget is not

---

[15]See Samuels, op. cit., chapter 8, for a more detailed analysis of the FARC project. Information about the Key Technology Center projects here is based on Samuels' more extensive analysis.

[16]David C. Mowery, *Alliance Politics and Economics: Multinational Joint Ventures in Commercial Aircraft* (Cambridge, Mass.: Ballinger Publishing Company, 1987), pp. 91-92.

[17]Foreign firms participating are United Technologies, GE, Rolls Royce, and Snecma.

[18]The U.S. Department of State approves export licenses for technology transfer by participating U.S. companies.

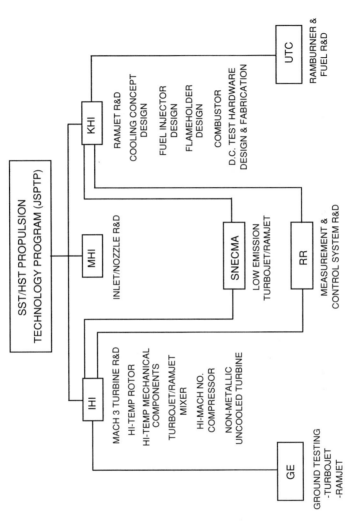

FIGURE 2-1 HYPR Technical R&D Responsibilities. SOURCE: Pratt & Whitney.

TABLE 2-7 NASA Aeronautics Programs (million dollars)

Category	1993	1994
Research and development	717	957
Aeronautics	(555)	(877)
National Aerospace Plane	(4)	(80)
Research operations support	149	144
Research and program management[a]	303	318
Construction of facilities	65	212
Total	1,234	1,631

[a]Includes aeronautics and national aerospace plane portions.
SOURCE: National Aeronautics and Space Administration.

TABLE 2-8 U.S. Government Outlays for Aeronautics R&D (million dollars)

Agency	1991
NASA (includes research and program management)	1,017
Department of Defense	6,792
Department of Transportation (FAA)	1,870
Total	9,679

NOTE: NASA figure includes research and development, construction of facilities, research and program management. Department of Defense figure includes research, development, and test and evaluation of aircraft and related equipment. Federal Aviation Administration figure includes research, engineering, and development; and facilities, engineering and development.
SOURCE: National Aeronautics and Space Administration, "Aeronautics and Space Report of the President" (annual), appearing in Aerospace Industries Association, *Aerospace Facts and Figures 1992-1993* (Washington, D.C.: AIA, 1992).

commensurate with the importance the industry plays in the nation's economy." The NRC committee recommended that NASA review its budget and emphasize the development of technologies that will make U.S. aeronautical products more competitive.[19] NASA's total budget for aeronautical R&D was $574 million in 1992; about 17 percent went to R&D contracted with industry.

Although there have clearly been cases where NASA-supported programs have produced technological advances that have enhanced the competitiveness

[19]National Research Council Aeronautics and Space Engineering Board, *Aeronautical Technologies for the 21st Century* (Washington, D.C.: National Academy Press, 1992), p. 7.

of U.S. firms,[20] *there has been no comprehensive effort directed toward technology commercialization and product application technology.* While some identify NASA's civil aeronautics program with industrial policy, there is growing interest today in coupling NASA's R&D more closely to industry, a theme that NASA took up in 1993.[21] Although Japan's National Aerospace Laboratory is funded at an annual level of about $100 million or less than one-fifth of NASA's budget for aeronautical R&D,[22] it does support some work in areas such as composite materials important to the future commercial aircraft industry. *Japan's government-supported domestic cooperative programs, particularly those supported by MITI, are more strongly oriented to technology sharing among Japanese companies and commercialization of technologies for commercial aircraft than those supported by the U.S. government* (see Tables 2-9 and 2-10).

The U.S. government has, however, played a major role in encouraging the development of air transportation, making the United States a leading market. This role constitutes an important source of indirect support for U.S. aircraft manufacturers. The main channels of support have been the Federal Aviation Administration (FAA) activities to ensure safety and to develop the air traffic infrastructure, and regulation of fares and routes by the Civil Aeronautics Board prior to its abolishment in 1978 with deregulation. With deregulation, the aircraft manufacturers lost the advantages of cooperation with deep-pocketed lead users (the airlines) who articulated demand and pushed product development. Although virtually all analysts agree that travelers have benefited from lower fares in the post-1978 period as increased competition has led airlines to reduce costs, current convulsions and heavy financial losses in the airline industry have caused some concerns about instability in the industry and raised doubts about the prospects for adequate long-term profitability.

TABLE 2-9 United States Aerospace Industry R&D Spending (million dollars)

	1988	1989	1990
Total	25,900	25,638	25,357
Federal Source	19,877	19,633	19,217
Industry Source	6,023	6,005	6,140

SOURCE: Aerospace Industries Association, *Aerospace Facts and Figures 1992-1993* (Washington, D.C.: AIA, 1992), p. 105.

---

[20]OTA, op. cit., p. 347.
[21]Kathy Sawyer, "Reviving Aeronautics—Space Agency Focuses on Global Context," *The Washington Post*, May 27, 1993, p. A23.

TABLE 2-10 Japanese Industry's Intramural Aircraft-Related R&D Spending[a] (million dollars, ¥110 per dollar)

Industry Sector	Aircraft-Related R&D Spending (% of total)			
	1990		1991	
Total[b]	442.6	(100%)	525.8	(100%)
Autos	82.1	(19%)	67.9	(13%)
Other transportation equipment	298.7	(68%)	408.7	(78%)
Aircraft and Parts	[18.2	(4%)]	[21.8	(4%)]
Other industries[c]	61.8	(14%)	49.2	(9%)

[a]Includes government funds spent by industry. [b]Total may not be exact due to rounding.
[c]Other industries conducting aircraft-related R&D during 1990 and 1991, none of which constituted more than 5 percent of the total, were textiles, chemicals, plastic products, rubber products, steel, nonferrous metals, machinery, electronic machinery, precision machinery, other manufacturing, and transportation/telecommunications/utilities.

SOURCE: Somucho Tokeikyoku (Management and Coordination Agency, Statistics Bureau), *Kagaku Gijutsu Kenkyu Chosa Hokoku—Heisei Sannen, (Report on the Survey of Research and Development 1991)*, (Tokyo: Nihon Tokei Kyokai, 1992), pp. 162-163; Somucho Tokeikyoku (Management and Coordination Agency, Statistics Bureau), *Kagaku Gijutsu Kenkyu Chosa Hokoku—Heisei Yonen (Report on the Survey of Research and Development 1992)*, (Tokyo: Nihon Tokei Kyokai, 1993), pp. 162-163; and Communication from the Management and Coordination Agency, Statistics Bureau, September 2, 1993.

U.S. government financing of aircraft exports at low interest rates through the U.S. Export-Import Bank provided strong support for the aircraft manufacturers in the 1970s. "Wars" over export financing were mitigated by the Large Aircraft Sector Understanding of the late 1970s, which set floors on acceptable interest rates. New financing techniques have, moreover, made private borrowing more feasible to purchase aircraft. In Japan, the Japan Development Bank and the Export-Import Bank continue to support aircraft imports with loans totalling $1.9 billion appropriated in 1992.[23] In recent years, The U.S. Export-Import Bank has again become important to aircraft exports (see Tables 2-11and 2-12).[24] Some have called for an increase in its budget for this purpose in order to address the current aircraft sales slump. Aircraft exports fell 15 percent in the first quarter of 1993, to $9.6 billion.[25]

[22]See NRC, op. cit., p. 8; and National Aerospace Laboratory 1991-1992 (program brochure), p. 4. The $100 million budget includes personnel as well as research and facilities for space and aircraft R&D.

[23] See Nihon Kokuchukogyokai, op. cit., p. 426.

[24]In August 1993, it was reported that the Export-Import Bank would provide loan guarantees for sales of aircraft to Saudi Arabia valued at more than $6 billion. See John Mintz and Ruth Marcus, "Saudis Shift Jetliner Order to U.S.," *The Washington Post*, August 20, 1993, p. B1.

[25]AIA, news release, June 16, 1993.

TABLE 2-11 U.S. Export-Import Bank (million dollars—1991)

Total loan authorizations	604
Loan authorizations supporting commercial jets	0
Total guarantee authorizations	6,016
Guarantee authorizations supporting commercial jets	566

NOTE: Commercial jet category includes complete aircraft, engines, parts, and retrofits.

SOURCE: Aerospace Industries Association, *Aerospace Facts and Figures 1992-1993* (Washington, D.C.: AIA, 1992), p. 134.

TABLE 2-12 U.S. Export-Import Bank 1992 Guarantees Supporting Commercial Jets (million dollars)

Country	Number	Type	Guarantee
Brazil	2	B-737	42.3
Mexico	1	B-737	30.4
Tanzania	2	B-737	52.8
Morocco	4	B-737	114.1
Chad	5	B-737	122.6
India	4	B-747	600.0
Norway	2	B-737	42.3
Pakistan	1	B-737	30.0
China	1	MD-11	94.5
Australia	5	B-737	130.7
Poland	9	B-737	246.1
China	1	MD-11	91.3
Total	37		1,597.1
(Export Value)			(1,889.1)
Total Guarantee Authorizations			7,301

SOURCE: Aerospace Industries Association and U.S. Export-Import Bank.

Like the United States, Japan is a signatory of the General Agreement on Tariffs and Trade and imposes no formal quotas on aircraft imports or formal offset requirements to increase Japanese-supplied content. Japan's three major airlines are now all formally private entities. However, U.S. manufacturers selling to Japan do feel informal pressures to source some parts in Japan in order to enhance access to Japanese airlines. The Ministry of Transportation has a major influence on the industry through its regulation of routes and fares.

The Treaty of Mutual Security and Assistance with the United States, Japan's only formal security treaty, is the bedrock of the defense relationship. This 40-year-old treaty[26] remains critically important to the overall bilateral relationship, although there is also a growing belief in the United States that the nature of the alliance will undergo change. The combination of threat reduction in Asia stemming from the end of the Cold War and budgetary pressures in the United States suggests that U.S. troop deployments in Asia will continue to decline, a prospect that worries some Japanese and other Asian allies. However, President Clinton's stress on the commitment of the United States to active engagement in the region and to multilateral discussions on security was well received in Japan.[27]

Military R&D and procurement constitute an area where U.S. and Japanese policies differ markedly. The U.S. Department of Defense (DOD) alone spent almost $7 billion in research, development, construction of facilities, and program management on aeronautics R&D in 1991 (see Figures 2-2 and 2-3).[28] Japan's defense budget, compared to that of the United States, allocates a smaller share to R&D. In 1992, for example, the ratio of capital equipment expenditure (including weapons procurement) in JDA's budget was 31 percent as contrasted to 2.5 percent for R&D (see Table 2-13). In the United States, the federal aeronautics budget for R&D was $9.6 billion and the total defense budget $273 billion.[29] Despite the fact that JDA's direct R&D funding is small, however, TRDI (JDA's Technical Research and Development Institute) focuses this effort on technologies that contribute to the overall industrial base. For example, emphasis on radar development and composite materials reflects an assessment that these technologies will have wide applications in both nondefense and defense areas.[30] In contrast, the DOD budget has focused increasingly in the last 15 years on areas such as stealth technologies that have no immediate applications to commercial aircraft. On the one hand, a higher percentage of aircraft production directed to military demand in Japan as compared to the United States suggests a strong effect on capital equipment spending by JDA. On the other hand, Japanese companies finance a large share of their R&D investments with their own funds, with the expectation of large, lucrative JDA procurement down the line.

In the early postwar period, as mentioned earlier, Japan's military aircraft industry was reborn on the basis of production licenses from U.S. firms, negotiated with the support of the U.S. government. Since the 1970s, the United

---

[26]The treaty was modified in 1960.

[27]See, for example, Ruth Marcus, "Summit a Winner for Clinton," *The Washington Post*, July 10, 1993, p. A1.

[28]AIA, *Aerospace Facts and Figures 1992-1993* (Washington, D.C.: AIA, 1992), p. 108.

[29]Outlays by NASA, DOD, and the Department of Transportation. Ibid. pp. 18 and 108.

[30]Michael W. Chinworth, *Inside Japan's Defense: Technology, Economics and Strategy* [Washington, D.C.: Brassey's (U.S.), 1992], pp. 42-44.

27

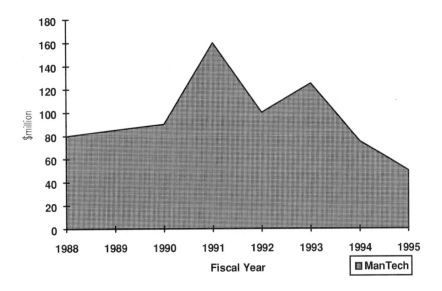

FIGURE 2-2 DOD manufacturing technology—fixed-wing aircraft. NOTE: ManTech shown for fixed-wing is about 50 percent of total ManTech. SOURCE: U.S. Department of Defense.

Fixed-Wing Aircraft

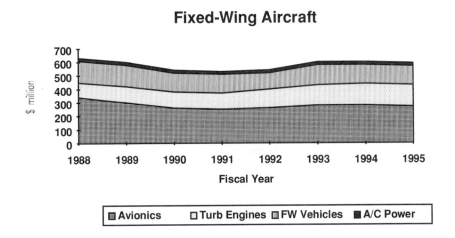

FIGURE 2-3 Aeronautical core R&D funding—fixed-wing aircraft. NOTE: Tech demos, dem/vals not included. Avionics include sensors, ASW, and EW technologies. SOURCE: U.S. Department of Defense.

TABLE 2-13 Japan's Defense Budget for Fiscal 1993 (million dollars, ¥110 per dollar)

	Amount	Percentage of Total
Personnel, provisions	17,632	41.8
Materiel	24,552	58.2
Equipment	9,811	23.3
R&D	1,125	2.7
Facilities	1,821	4.3
Maintenance	6,855	16.3
Base countermeasure costs	4,401	10.4
Other	540	1.3
Total	42,187	100

SOURCE: Boeicho (Japan Defense Agency), ed., *Heisei Gonen Boei Hakusho* (1993 Defense White Paper), (Tokyo: Okurasho Insatsukyoku, 1993), p. 333.

States has attempted to increase the flow of defense-related technology from Japan. Despite the 1983 exchange of notes in which Japan agreed to exempt the United States from its political prohibition on military technology exports, and the work of the U.S.-Japan Systems and Technology Forum, the results have fallen far short of expectations. There are a number of possible explanations, including a lack of understanding about what the United States wants from Japan, as well as Japanese reluctance to transfer technologies to the United States that might be incorporated into weapons systems and retransferred abroad.

Japan has strongly emphasized dual-use facilities in its defense R&D. Indeed, Japan's world-class commercial industrial base is seen as the foundation for military production. A former director of the TRDI, JDA's R&D institute, has noted that in technology there is no black or white, only gray—it becomes military or civilian in application.[31] Japan's approach to military R&D has been not to focus on technology breakthroughs, but rather to stimulate industrial sectors and technologies that have a wide range of applications, carefully arranging for a division of labor among companies that promotes building of specialized skills that complement those of other firms. Japanese companies have developed substitute components for weapons systems licensed from the United States (either because the components were "black boxed" and Japan wished to develop independent technology or as improvements on U.S.-origin technologies). These components are commonly derived from commercial products, often without Japanese government funding. In general, the Japanese consider these technologies to be nonderived and interpret provisions for exemption of commercial technologies from technology flowback arrangements with the United States quite broadly.

[31]Ibid., p. 36.

Although there are sharp contrasts in the nature of military-civilian interface in the two countries, the committee concludes that Japan does a better job of effectively utilizing its resources to promote synergies between military and civilian aircraft production. DOD procurement practices pose significant obstacles to companies that wish to promote military-civilian synergies. Accounting practices, military specifications, unique contract requirements, and policies on technical data rights all inhibit interactions and force companies to use separate plant facilities.[32] The large amount that DOD spends on R&D compared with the Japanese government clearly benefits U.S. industry in supporting its technology base, but the synergy between U.S. military and commercial technology has been declining.[33] In contrast to DOD's support in years past for technologies (jet engines and swept back airfoils) with both military and defense applications, during the past 15 years, DOD has oriented its support to defense-unique technologies such as stealth and high maneuverability.

In Japan, while there also exist some obstacles to military-civilian interactions related to military specifications and procurement practices, there are offsetting factors. The fact that military aircraft production is carried out by large Japanese companies with diversified production in other sectors, as well as the colocation of military and civilian production lines create opportunities for cross-fertilization of manufacturing know-how and sensitivity to the potential applications of technologies developed on the commercial side. Japan's procurement system helps to reinforce "technology highways" that link larger companies with suppliers, integrate military and civilian production, and foster an integrated and flexible dual-use technology and industrial base. In Japan, technological and commercial competence is as much a matter of national security as force deployment.[34]

Dramatic cuts in the U.S. defense budget in recent years have resulted in a fundamental restructuring within the industry and companies engaged in military production are pursuing a combination of downsizing, consolidation, diversification, and exit strategies. In Japan, industry observers are also worried about declining defense procurement, which is expected to hit the industry hard in the mid-1990s.[35] The push toward commercial production is thus a clear im-

---

[32]Report of the Center for Strategic and International Studies Steering Committee on Security and Technology, *Integrating Commercial and Military Technologies for Military Strength* (Washington, D.C.: CSIS, 1991).

[33]See John A. Alic, Lewis M. Branscomb, Harvey Brooks, Ashton B. Carter, and Gerald L. Epstein, *Beyond Spinoff: Military and Commercial Technologies in a Changing World* (Boston, Mass.: Harvard Business School Press, 1992).

[34]For a detailed analysis of Japan's "technology highways" and the "protocol system" among companies, see Samuels, op. cit., chapter 8; and David B. Friedman and Richard J. Samuels, "How to Succeed Without Really Flying: The Japanese Aircraft Industry and Japan's Technology Ideology," in J. Frankel and M. Kahler, eds., *Regionalism and Rivalry: Japan and the U.S. in Pacific Asia* (University of Chicago Press, 1993).

[35]In FY 1991, JDA's defense acquisition budget was cut 16.1 percent over the previous year. The Air Self-Defense Force received funds for 29 F-15 fighters rather than 42. Procurement of four AWACS ($465 million each)—which are produced in the United States by Boeing—is planned during the current

perative in both countries. One key question is whether Japan's aircraft industry may be particularly well positioned to capture increasing shares of the aircraft and commercial engine components manufacturing, and repair markets in the future. Based on examination of the policies (public and private) that have fostered close integration of large and small companies, flexibility of capital equipment, and tight coupling of defense and commercial production, the committee judges it likely that the already apparent trends of increasing Japanese shares in these areas, particularly components manufacturing and repair markets, will continue in the future.

Planners in both Japan and the United States, attempting to adjust to the dramatic changes mentioned at the outset, are considering new approaches. In the United States, the Advanced Research Projects Agency (ARPA) is leading a new set of programs aimed at fostering defense conversion, while DOD's leadership is focusing on reducing barriers between military and civilian production through streamlined procurement in the context of a lower defense budget, and NASA has announced a new stress on aeronautical R&D. Meanwhile, a commission on the future of the airline industry has recommended policy changes relevant to that industry.[36] In the United States, the approach to policy redirection appears to be largely ad hoc and uncoordinated, whereas Japan's decision-making agencies are fewer in number and work together to formulate a common vision for the industry.

These differences have significant implications for U.S. and Japanese companies interested in forming partnerships. *A Japanese company interested in forming a technology linkage with a potential U.S. partner coordinates with a smaller number of key actors in government than does the U.S. company* (see Table 2-14).

Within the government, MITI is the major player, but interactions with JDA are also required with respect to military programs. MITI's Aircraft and Ordnance Division, which plays the central role in policy formulation, has shifted policy focus on "national production" (*kokusanka*) to international joint ventures.[37] Japan's major aircraft companies and suppliers are members of the Society of Japanese Aerospace Companies (SJAC), which sometimes acts as coordinator (as has been the case with the international consortium on commercial aircraft components and foreign missions such as the recent trip to Russia) and Keidanren's Defense Production Committee. In contrast to the situation in the United States, the number of actors is smaller, the major players

---

1991-1995 plan. Japanese defense planners worry that procurement of two in FY 1993 will account for a large share of JDA's total defense procurement budget for all services. See Barbara Wanner, "Japanese Defense Industry Grapples with Post-Cold War Conversion," *JEI Report*, No. 12A, April 2, 1993.

    [36]See National Commission to Ensure a Strong Competitive Airline Industry, *Change, Challenge and Competition: A Report to the President and Congress* (Washington, D.C.: U.S. Government Printing Office, 1993).

    [37]Michael Green, *Kokusanka: FSX and Japan's Search for Autonomous Defense Production* (MIT Japan Program Working Paper, 1990).

TABLE 2-14 Major Policy Checkpoints for Companies Forming International Technology Linkages

Government	Industry
Japan	
MITI	SJAC
Aircraft and Ordnance Division	
Aircraft Industry Council	
AIST	
JDA	Keidanren Defense
Equipment Bureau, Aircraft Division	Production Committee
Procurement Office	
TRDI	
Air Self-Defense Staff Tech Department	
STA	
National Aerospace Laboratory	
Related Organizations	
IADF	
United States[a]	
Department of Commerce	AIA
International Trade Administration,	American League for
Aerospace, Trade Development	Exports and Security
International Economic Policy, Japan	·Assistance
Bureau of Export Administration	
Industrial Resources Administration	
National Security Preparedness Division	
Technology Administration	
Department of Defense	
International Security Affairs, Japan Desk	
Defense Security Assistance Agency	
Acquisitions, International Programs	
ARPA	
Defense Technology Security Assistance Administration	
Military Services	
Department of State	
Bureau of East Asia-Pacific, Political Affairs	
Bureau of Political Military Affairs, National Security	
Defense Relations, Security Assistance	
Center for Defense Trade	
Bureau of Economic and Business Affairs	
U.S. Trade Representative	

[a]Consultations differ, depending on the program. The FS-X project involved consultations with most of these agencies.

SOURCE: Based on memos provided to the committee by Michael Green and Gregg Rubinstein.

overlap, and the process is oriented toward quiet advance coordination among business and government.

A U.S. company considering a technology linkage with a Japanese counterpart interacts with a more complex maze of U.S. agencies and regulations. In the case of a military project, the company must consult with a variety of offices within DOD, including the Japan desk of what is now Regional Security Affairs; the Defense Security Assistance Agency (DSAA), which coordinates defense sales and licensed production; and the Defense Technology Security Administration, which oversees technology transfer and export licenses for DOD.[38] All military export applications are submitted to and approved by the Department of State, Office of Defense Trade Control. For military or commercial projects, a U.S. company is well advised to consult with the Department of Commerce (the International Trade Adminstration and the Bureau of Export Administration) as well as the State Department.

Under normal procedures, approval of licenses is handled by the licensing offices of the various departments in coordination with the relevant program offices and country desks. However, official evaluation of military aircraft programs is complicated by the arbitrary division of responsibility in DOD for sales/licensed production and cooperative R&D programs in two separate and often uncoordinated bureaucratic entities (Undersecretary for Policy, DSAA and the Undersecretary for Acquisition, Dual-use Technology Policy & International Programs). This has often led to inconsistency in DOD positions on Japan programs, as well as problems in coordinating with other agencies. The increased role of the Department of Commerce in recent years reflects a recognition that the U.S. industrial/technology base is both a defense and an economic policy concern, but in practice, effective coordination among DOD, Commerce, and State is often difficult. In controversial cases, senior executive branch officials participate in an interagency process coordinated through the National Security Council or the National Economic Council and draw the attention of members of Congress and research organizations such as the General Accounting Office.[39]

In Japan, the process of policy evaluation and adjustment is also multifaceted, but the locus of activity is clear: MITI and the industry. Japanese industry and government became more realistic in the 1980s concerning obstacles to becoming a world-class player in aircraft; MITI can not and does not direct the industry, but develops policy jointly with industry. Compared to other sectors such as computers and semiconductors, which are also the focus of policy, MITI has considerable influence over the aerospace industry because industry is

---

[38]In addition, the Office of International Programs has jurisdiction relating to R&D programs (as does ARPA potentially), and consultations with the military services are essential for all cooperative projects involving military aircraft.

[39]This material is summarized from memos prepared for the committee by Michael Green and Gregg Rubinstein.

still highly dependent on JDA procurement, has not moved offshore, and is restricted from defense exports.

Japanese government and industry continue to look ahead to the future, planning new programs and policy adjustments. In past months, a series of new studies, working groups and international missions have been organized to consider critical decisions relevant to the future of the industry (see Table 2-15). For example, MITI and the Ministry of Transportation in cooperation with SJAC have formed a committee to study requirements for the High Speed Commercial Transport (HSCT). The purpose is reported as developing a "Japanese proposal" for presentation to Boeing and Airbus concerning future specifications and domestic infrastructure requirements. Meanwhile, MITI, JDA, and SJAC are reportedly formulating a domestic development program for a medium-sized transport that can be used for military and commercial purposes. JDA has set up two working groups to look at defense procurement and R&D activities in defense technology. Some of this work, such as the HSCT study, will be made public. Other activities, such as the JDA working groups, will continue discussions for a number of months with no expectation of producing published reports.

International linkages are very much a focus of planning. SJAC recently sent a mission to Russia, with a resulting plan to invite Russian engine specialists to Japan and expand access of Japanese companies to Russian test facilities.[40] Airbus has, meanwhile, expressed interest in cooperating with Japan's committee examining HSCT issues. In the context of the U.S.-Japan Systems and Technology Forum, one new cooperative project on ducted rocket engine technology was initiated and others are in the planning stage. As the development stage is completed on the FS-X, it is expected that negotiations will begin on production.

All of these efforts will feed into a process that provides Japan with the option for aircraft production in the twenty-first century. Many of the same individuals are key participants in all of the Japanese studies and missions. It may be some time before a change in Japan's official policy is formally articulated. In the meantime, a process of information gathering, foreign travel, discussion, and exchange will take place that builds a common framework for making choices. In this process, industry and government interact as partners who share a common overarching goal.

Japan has more alternatives for international partnerships than ever before in the postwar period. With whom and how to form linkages of various sorts are major considerations. Increasingly, Japanese companies are experimenting with diverse partnerships that involve more than one foreign company. Further diversification of international linkages seems likely, but geopolitical questions remain, such as whether Japan and Russia can resolve their lingering World

[40]U.S. firms are also expanding their linkages to Russia. For example, Pratt & Whitney will supply engines and Collins will supply avionics for the new Ilyushin IL-96M aircraft, which reportedly will sell at a cost far below similar sized airplanes now on the market.

TABLE 2-15 Aircraft Industry-Related Studies in Japan

1. Requirements for HSCT and Very Large Transport
   MITI and the Ministry of Transportation, in cooperation with SJAC, are forming a "Committee to Promote the Introduction of Next-Generation Aircraft." Including representation from the four heavies and the three largest airlines, the committee will study demand for the superjumbo and HSCT. The purpose will be twofold: (1) to present a "Japanese proposal" to airframe manufacturers concerning the specifications of these future aircraft; and (2) to study the domestic infrastructure implications of introducing them (*Nihon Keizai Shimbun*, April 13, 1993).

2. SJAC-Russia Joint Programs
   A mid-May 1993 SJAC mission to Russia resulted in an agreement to invite Russian engine specialists to Japan and for Japanese companies to gain access to Russian test facilities (*Japan Digest*, May 27, 1993).

3. Multipurpose Medium Aircraft
   A joint planning committee of MITI, JDA, and SJAC has reportedly been charged with formulating a domestic development program for a medium-sized transport that could be used by domestic airlines and by JDA for transport and antisubmarine roles. Total production volume is anticipated to be 300-500 (*Nikkan Kogyo Shimbun*, January 19, 1993).

4. Second-Generation SST Studies
   Since 1987, the Supersonic Transport Development and Survey Committee of SJAC has conducted studies under commission from MITI on the airframe specifications for next-generation SSTs, so that key technologies could be identified and developed (*Kokusai Koku Uchu*, December 1992).

5. YS-X Transport
   MITI funding continues for research on the 75-100 twin-turbofan. Japan would take the lead in an international partnership (*Aviation Week and Space Technology*, June 1, 1992).

6. YXX/7J7 Transport
   MITI funding for this 100+ seat transport has continued, although future prospects are uncertain.

7. Study on the Future of the Japanese Aircraft Industry
   A MITI-led study was mentioned by Keidanren Defense Production Committee during a July 25, 1993 meeting with the NRC committee.

8. Basic Technology for Advanced Stealth Aircraft
   TRDI is reportedly proposing work on a proof-of-concept aircraft to begin as FS-X and OH-X development winds down in 1995 (*Aerospace Japan-Weekly*, June 14, 1993).

9. C-X Transport and T-X Trainer
   These are indigenous aircraft programs reportedly being considered by JDA. Connection between C-X and dual-use transport (item 3 above) is unclear (*Aerospace Japan-Weekly*, June 14, 1993).

10. Test Facilities
   Planning continues for new aircraft and rocket engine testing facilities under the auspices of TRDI (*Aviation Week and Space Technology*, August 24, 1992).

11. International Composites Program
   Press reports during the summer of 1992 described a new MITI program researching applications of lightweight composite materials for supersonic aircraft. The proposed program would run for six years, cost $240 million, and be open to foreign participation (*Aviation Week and Space Technology*, August 3, 1992).

TABLE 2-15 *Continued*

12. JDA Advisory Committees

JDA's Equipment bureau has reportedly formed two advisory committees that do not appear to be connected with any particular potential program on Defense Equipment Procurement and on Defense Industry Technology. (Source: Mutual Defense Assistance Office.)

SOURCE: Compiled by National Research Council Committee on U.S.-Japan Aircraft Linkages from various sources.

War II era territorial dispute over the Northern Territories. Another consideration is whether Boeing will pursue an ever-broadening and deepening role for Japanese companies. Airbus has been exploring cooperation with Japan, with success seen in expanded sales of aircraft in recent years.

The Japanese policy and business environment allows industry to gain maximum leverage from international alliances and procurements, resulting in a gradual upgrading of independent technological capabilities and diffusion of those skills across civilian and military production and among the major contractors and the many subcontractors in Japan's aircraft manufacturing network. The Japanese aircraft industry does not carry out full independent integration of airframes, but it has become a major player in the subsystems and components areas and, with the support of the government, has built significant indigenous capabilities. Japan has achieved increasing independence and growing technological strength by promoting international linkages, particularly in the defense area.[41] Japan is pursuing international linkages and the development of indigenous capabilities simultaneously, skillfully managing international cooperation to derive maximum gains in terms of autonomous development.

---

[41]See Samuels, op. cit., chapter 8 for an analysis of "the paradox of autonomy through dependence." Samuels outlines how technology agreements permit the accumulation of skills with broad competitive implications. In this process, the government of Japan has played a strong role in managing competition and providing incentives for cooperative activities.

3

Current Status Of U.S.-Japan Linkages

Drawing on published information, briefings from experts, and its study mission to Japan, the committee examined a wide range of U.S.-Japan technology linkages relevant to transport aircraft. The assessment included prime program partnerships and government-supported R&D programs as well as relationships at various levels of the supplier chain. This chapter summarizes the information on linkages the committee has collected; analyzes the motivations, mechanisms, and impacts of linkages; and highlights major themes and insights. More detailed materials on linkages are contained in Appendixes B and C.

AIRFRAMES

Linkages in Commercial Airframes

The most significant U.S.-Japan linkages in the commercial airframe segment are the series of program-based alliances concluded between Boeing and the Japanese "heavies."[1] To all accounts, this relationship has brought significant benefits to both sides.

From the start of the 747 program in the late 1960s through the subsequent 737 and 757 programs, Boeing procured parts and equipment from Mitsubishi Heavy Industries (MHI), Kawasaki Heavy Industries (KHI), and Fuji Heavy Industries (FHI). Starting with the 767 program in the late 1970s and continuing with the 777—which is scheduled to enter service in 1995—the Boeing-Japan interaction has evolved from one in which the Japanese companies "built parts to specification" to actual design and engineering interaction from the earliest stages of product development. The work share and the technical sophistication of the manufacturing tasks undertaken by the Japanese partners have also increased steadily over time.

Boeing's primary motivation for approaching the Japanese heavies about significant participation in the 767 program was the perception that the linkage might bring market leverage. The Japanese were probably most motivated by a desire to gain access to technology as well as indirect access to the global aircraft market. MHI, KHI, and FHI designed and now manufacture approximately 15 percent of the airframe of the 767, a wide-body twinjet. As "risk-sharing subcontractors," the Japanese partners assumed the risk for their non-equity share in the program, including tooling and other investment. The Japanese government provided funding through success-conditional loans for much of this investment.

Boeing, the three heavy industry companies, and the Japanese government through the JADC negotiated a "program partnership" for the subsequent 777 program. This alliance is similar to the 767 arrangement, although Boeing originally offered the Japanese partners significant program equity participation, which they were not willing to assume. The Japanese work share in the 777 program is higher than in the 767—Japanese partners essentially build all the fuselage parts except for the nose section, as well as the wing center section, the wing-to-body fairing, and landing gear doors.[2] Indirect Japanese government support and Japan Development Bank loans have also been made available to the heavies for their participation in the 777 program.

Japanese technical responsibilities increased with the 777. There were many more Japanese engineers involved in 777 development than in 767 development, with several hundred sent to Seattle during the most intensive design phase. As was the case in the 767, the Japanese are limited in the engineering effort to their own work package.

---

[1]There are three other linkages of note: (1) the Japanese heavies manufacture some components for McDonnell Douglas; (2) Mitsui & Co., McDonnell Douglas's trading company, played a key role in financing the launch of the MD-11 (more a business alliance than a technology linkage); and (3) the Toyota-affiliated Ishida Group has made several direct investments in small U.S. companies, including an undertaking to develop a tilt-rotor aircraft, which was reportedly suspended earlier this year.

[2]Japanese companies also build the wing-to-body fairing on the 767.

On both the 767 and the 777 programs the direction of technology transfer was predominantly from Boeing to its Japanese partners. This took several forms, including data exchange and engineer training in the use of advanced computer design techniques. Boeing limited the transfer of its critical technologies by keeping to itself the design and manufacture of the most sensitive parts of the airframe as well as all the systems integration activities. Boeing also implemented management systems that allow engineering data exchange to be managed on a "need-to-know" basis. Some technology also flows to Boeing from Japanese companies, particularly approaches to manufacturing technology and processes.

In addition to the 767 and 777 partnerships, Boeing collaborated with the Japanese heavies on preliminary design and market definition work for a proposed 150-seat transport—the 7J7-YXX program. This program contemplated significant Japanese equity participation and interaction in areas such as marketing that the 767 and 777 partnerships did not encompass. Although the Japanese government still supports work related to the YXX, program launch has been put on hold.

The Boeing-Japan relationship appears to have delivered significant benefits to both sides that roughly parallel their likely initial motivations. In addition to aircraft sales in the Japanese market, the program partnerships have allowed Boeing to spread a significant part of the program financing load. To this point, the Japanese heavies have not entered partnerships with the other two major airframe manufacturers, and have not emerged as a significant competitive threat to Boeing. Boeing has also gained access to competitively priced, high-quality components.

For the Japanese heavies, the Boeing alliance has delivered technology and know-how, a significant stream of long-term business, relatively low-risk access to global aircraft markets, and government support in developing their technology and manufacturing bases. The Japanese participants have also hit some rough spots along the way. For example, exchange rate shifts during the 1980s and more recently, as well as the current market downturn, have made apparent some of the liabilities associated with risk sharing.

Perhaps most importantly, the Japanese heavies have developed a world-class manufacturing infrastructure and technology base for aircraft structures. This capability—largely built in conjunction with their work on Boeing programs—has implications for U.S. structures suppliers.

Japanese Capabilities in Structures Manufacture and Implications for U.S. Suppliers

As described above, a major focus of Japanese industry in the production of commercial transports is in the area of structures, particularly supplying Boeing on the 767 and 777 programs. Figure 3-1 shows the global players in this area of aircraft manufacturing, broken down according to the parts of the aircraft

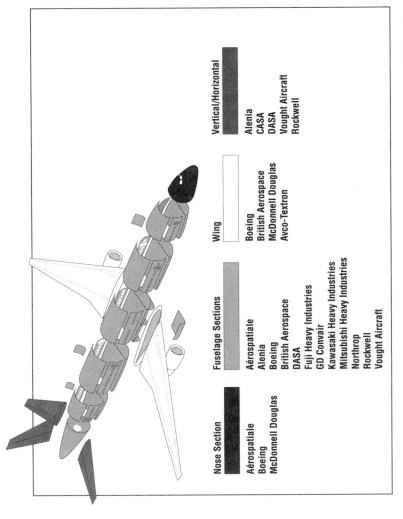

Nose Section

Aérospatiale
Boeing
McDonnell Douglas

Fuselage Sections

Aérospatiale
Alenia
Boeing
British Aerospace
DASA
Fuji Heavy Industries
GD Convair
Kawasaki Heavy Industries
Mitsubishi Heavy Industries
Northrop
Rockwell
Vought Aircraft

Wing

Boeing
British Aerospace
McDonnell Douglas
Avco-Textron

Vertical/Horizontal

Alenia
CASA
DASA
Vought Aircraft
Rockwell

FIGURE 3-1 Commercial airframe manufacturers. SOURCE: National Research Council Working Group on U.S.-Japan Technology in Transport Aircraft.

manufactured. The figure is not exhaustive, and it focuses on structure suppliers for large commercial transports—particularly wide bodies. The airframe "primes" tend to retain manufacture of the wing (excluding control surfaces) and the nose section, the latter primarily because of its importance for integration activities as the "brain" of the aircraft.[3]

The basic manufacturing process for fuselage parts involves considerable subassembly. Premium aluminum skins are attached to aluminum "stringers" in order to create skin panel subassemblies. These panels are then attached to each other with large fuselage frames to form larger fuselage segment subassemblies, a complementary set of which is fitted together to form a hollow "barrel" section assembly. The barrels are then either "stuffed" with subsystems (i.e., electronics, hydraulic, and environmental systems), before being joined or joined into larger sections before being stuffed.

Various considerations, such as transportation, affect the manufacturing process. In the case of Airbus, for example, the fuselage sections manufactured by member companies are stuffed before being shipped to Toulouse, France, where they are joined together. This is similar to the process for some U.S. military programs such as the F/A-18, in which Northrop stuffs and tests sections before shipping them to McDonnell Douglas. In the case of the Boeing 767 and 777, the Japanese heavies ship the fuselage panels to Boeing, as Northrop, Rockwell, and Vought do for the 747, and Boeing assembles and stuffs the sections.

A number of factors—such as capital availability—influence the introduction of new technology into these processes, and some companies are more aggressive than others in applying new technology. The committee was very impressed with the technology level and breadth of the structures manufacturing capability possessed by the Japanese heavies. *Perhaps the most striking aspect of this capability is the advances the heavies have made in combining technologies transferred from the United States with the world-class manufacturing practices widely followed in other Japanese industries to create new process technologies.*

This is apparent in Japanese innovations in the skin panel process. Figure 3-2 shows estimated Japanese technical milestones in airframe structures. Some of the technologies, such as CATIA[4] computer-aided design software (CAD), were purchased by the Japanese heavies or were transferred from the United States through commercial and military programs. For example, by integrating the CATIA data base, which contains the hole locations for all variations of stringers used on the 777, with an automated drill, the heavies have worked with their machine tool suppliers and/or divisions to develop an automated universal stringer drill station. Different stringer variations can all be drilled on this station by reprogramming, thereby eliminating the need for specialized

[3] This situation is evolving. For example, Fuji is supplying the wing center section for the 777.

[4] CATIA (computer-aided, three-dimensional, interactive application) was first developed by Dassault and later improved by IBM and Boeing.

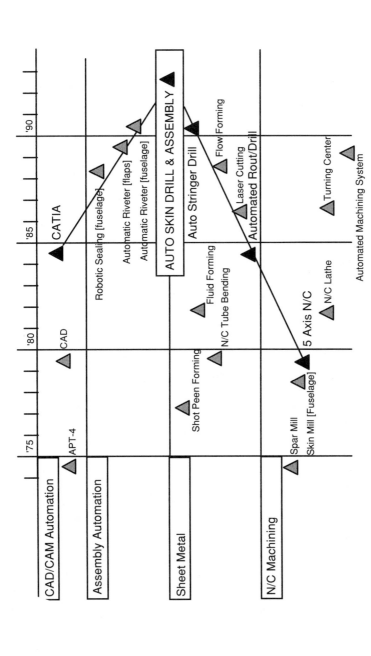

FIGURE 3-2 Japanese airframe structure technology milestones (estimated). SOURCE: National Research Council Working Group on U.S.-Japan Technology Linkages in Transport Aircraft.

tools. The Japanese structure makers are utilizing CATIA-controlled five-axis machines to automate other structures manufacturing steps, such as the chem-milling and drilling of aluminum skins. A further notable feature of Japanese structures capability is its breadth across the three heavies. For example, center wing section panels manufactured at Fuji use thicker aluminum skins than the fuselage panels manufactured at the other two heavies and require different manufacturing processes. Taken as a whole, these manufacturing technology improvements are good illustrations of the well-honed process of technology improvement and deployment that exists in many of the best Japanese manufacturers. Robotics and other new machinery are developed and deployed by the heavies as part of a system that maximizes the impact of new technology on the entire manufacturing process. These manufacturing practices are well known in the automobile and other mass production industries, but the application of new technology to aircraft production—which involves much smaller production runs—is perhaps more challenging because of the difficulties of achieving scale economies. Japan's aircraft makers do not utilize technology for its own sake, but focus on process improvements that deliver a competitive advantage in terms of cost and quality. Although the basic process for improving and combining technologies, as well as some of the constituent technologies, already existed in Japan and had been applied and proven in other industries, several of the key manufacturing capabilities were transferred from the United States.

What are the implications for U.S. structures makers, who are challenged by pressures on the defense sides of their businesses as well as the globalization of commercial structures procurement? First, it is necessary for American companies to stay abreast of developments in Japan and elsewhere. It is usually not difficult for Americans to at least tour the factories of the Japanese heavies and see their manufacturing processes. Second, American companies must be aggressive in seeking to transfer Japanese technologies back to their operations. While companies such as Northrop, Vought, and Rockwell must focus their business strategies on future programs, they must also invest in new technologies and related equipment to remain competitive.

The Japanese example also shows that the challenges facing the American structures suppliers go beyond the imperative of monitoring and learning specific manufacturing innovations from Japanese competitors. In order to implement world-class manufacturing solutions that require large capital investments on the order of what the Japanese heavies have made, a significant business base is required. This can come only from participation in new programs, which is problematic for U.S. suppliers. Because American primes feel that procurement from Japanese and other foreign suppliers is a critical element in enabling sales in these markets (in some cases through formal offset requirements, or through informal signals and pressure in the case of Japan), U.S. structures suppliers must be particularly competitive in price, quality, and delivery performance to match the Japanese heavies or other international suppliers.

Linkages in Military Airframes

Perhaps the most extensive U.S.-Japan technology linkages in aircraft manufacturing and—more recently—in design have occurred in military programs. Licensed production, coproduction, and codevelopment of military aircraft undertaken in the context of the U.S.-Japan security alliance have resulted in a significant transfer of U.S. technology to Japan. The two most important U.S.-Japan military aircraft linkages in the recent past have been licensed production of McDonnell Douglas's F-15 and codevelopment of the FS-X.

Japanese companies had assembled the North American F-86 in the 1950s, and had produced the Lockheed F-104 and the McDonnell Douglas F-4 under license in the 1960s and 1970s. Japanese licensed production of the F-15 beginning around 1980 was an important step in the evolution of Japan's aircraft industry and U.S.-Japan defense technology relations. Although there were early national security concerns in the U.S. Department of Defense (DOD) over the transfer of advanced technology, the broad U.S. strategic and political rationale for Japanese production—primarily a greater contribution to regional security from a more militarily capable Japan—prevailed without a great deal of contention in the U.S. government.

The United States provided technologies and data necessary for Japanese production of the F-15, with the exception of a number of items such as design data, radar, electronic countermeasures, software, and source codes classified as "nonreleasable." The extent of this "black boxing" was greater than in the F-4 program and provided a motivation for Japanese industry to pursue the independent Japanese development of the country's next fighter in the mid-1980s. Still, the technology transfer was substantial in terms of quantity, and some argue that the level of technology transferred through F-15 licensed production was significantly higher than in previous bilateral programs.[5]

Soon after the launch of F-15 production, the Japan Defense Agency (JDA), ASDF, and Japanese industry began considering options for replacing the domestically-developed F-1 fighter. Industry and some elements in the government began the process with a presumption in favor of a domestically-developed fighter. Increasing domestic content, gaining greater managerial control over the program than was possible in a licensed production arrangement, and controlling costs were all considerations. Another important factor was an underlying sense that Japan's position in the aircraft industry was fragile and that

---

[5]"The initial list of technical data to be made available to the Japanese in the F-15 program, for example, consisted of 21 pages listing more than 300 items that in turn consisted of everything from single drawings and rolls of microfilm to magnetic tapes and boxes of microfiche." Michael W. Chinworth, *Inside Japan's Defense: Technology, Economics and Strategy* [Washington, D.C.: Brassey's (U.S.), 1992], p. 117.

passing up domestic development would consign Japan to a follower role forever.[6]

During 1986, by which time the momentum in Japan for domestic development had become quite strong, DOD began a more aggressive push for the FS-X to be based on an existing U.S. design. This resulted in an agreement to "codevelop" an FS-X based on the design of the General Dynamics F-16. From the start, the two countries conceived codevelopment differently, making it an attractive political solution but ensuring problems later. The Japanese assumed that a Japanese company would manage the process of developing an indigenous aircraft, with selected foreign technologies incorporated as necessary. The U.S. conceived the joint improvement of an existing aircraft, with a priority on ensuring "flowback" of Japanese technology based on know-how transferred by the United States.

A U.S.-Japan memorandum of understanding (MOU) on FS-X codevelopment was signed in late 1988, but congressional concerns were raised during confirmation hearings of Bush administration officials in early 1989. Contentious debate over the agreement continued through the spring of that year, with opponents arguing that F-16 technology transfers would contribute to Japanese competitiveness in commercial and military aircraft, that "off-the-shelf" Japanese procurement of F-16s would cut the huge U.S. trade deficit with Japan while addressing Japan's security needs more economically, and that Japanese technical capabilities were not high enough for the flowback provisions to deliver many benefits to the United States. U.S. proponents of FS-X codevelopment argued that significant U.S. participation in the FS-X program was better than none at all, that Japanese procurement of unmodified F-16s was not a realistic scenario, and that flowback would bring considerable benefits.

In the end, congressional opponents were not able to stop the FS-X agreement, but were able to force DOD to gain a "clarification" of several key points. First, the Japanese explicitly committed to a 40 percent U.S. work share during the development phase and to providing access to Japanese-developed technologies. Second, the denial of several key F-16 technologies—including computer source codes, software for the fly-by-wire flight control system, and other avionics software—was made explicit.

The clarification exercise threw into sharp relief the contrast between the contentious divisions over Japan policy in the United States and the much more united front—albeit with some bureaucratic infighting—that Japan presents to the United States in bilateral negotiations. In addition, the contention left heightened resentment on both sides. Many Japanese opinion leaders, in particular, resent codevelopment as having been forced on Japan by the United States.

The development phase is now nearing completion, and first flight is projected for September 1995. Prospects for actual procurement are still uncertain.

---

[6]Ibid., p. 138.

If the FS-X goes into production, negotiation of a U.S.-Japan production MOU could be complicated by lingering disagreements over classifying derived and nonderived technologies, and U.S. work share.

In assessing the impact of U.S.-Japan collaboration in military programs on the technological capability of Japan's aircraft industry, analysts present a mixed picture. There is general agreement that Japanese companies receiving technology through F-15 licensed production were in a better position to supply the subsequent FS-X program. Impacts on the commercial side are less clear. At the supplier level, although a large number of Japanese suppliers make similar components for the F-15 and for the Boeing 777, many of these companies were supplying Boeing programs prior to the 777.[7] Still, the importance of military work (which accounted for more than 73 percent of Japan's total aircraft industrial output in 1990) for Japan's aircraft manufacturing and technological capabilities should not be underestimated. For example, Ishikawajima-Harima Heavy Industries (IHI) developed the capability to manufacture the long shafts for aircraft engines through the F100 program (described below in the section on engine linkages) and has evolved into a global center of excellence for this component. In addition to supporting specific dual-use technologies, Japanese military procurement supports equipment spending and engineering employment that are available for utilization on the commercial side.

At the prime level, analysts have pointed out that the FS-X program is structured to develop systems integration skills—a major missing piece of the puzzle for Japan's overall capability in aircraft. Although source codes and other critical items were not transferred, the considerable modification of the F-16 necessitated the transfer of design and systems integration technology from the United States to Japan—a first in bilateral military programs. The extent to which the Japanese will be able to capitalize on this technology in the future—in military as well as commercial aircraft development—is still an open question. There is, however, no question that it is a help.

There is also considerable disagreement about the value of Japanese technology developed through the FS-X program to which U.S. industry will have access to (either as flowback or through licensing). According to some reports, Lockheed (which purchased the Fort Worth fighter division from General Dynamics in 1992) has found the flowback of composite wing technology from Mitsubishi to be useful.[8] At this point, however, data are not being disseminated widely to U.S. industry, and some experts assert that a more systematic effort is needed to assess the value of FS-X technology flowback.

In the area of composites, the committee saw an interesting contrast between U.S. and Japanese systems of civil-military aircraft technology integra-

---

[7]U.S. General Accounting Office, "Technology Transfer: Japanese Firms Involved in F-15 Coproduction and Civil Aircraft Programs," GAO/NSIAD-92-178, June 1992.

[8]Alan S. Brown, "What Can Japan Teach the U.S. About Composites," *Aerospace America*, July 1993, pp. 36-40.

tion. These differences have significant implications for U.S.-Japan linkages in a critical area of future aircraft technology development.

COMPOSITES

Japanese Capabilities and U.S.-Japan Linkages in Composites

As described above, most aircraft structures are made of aluminum and have been for more than fifty years. U.S. companies, most notably Alcoa, are leaders in producing the high quality aluminum used in aircraft and aerospace applications, holding more than 80 percent of the world commercial transport market excluding the former Soviet Union and China.

Although the U.S. position in aluminum is strong—a new alloy developed by Alcoa has been specified for use on the Boeing 777—composite materials have been gradually incorporated into airframe structures over the past two decades. They possess several properties—mainly higher specific strength and lower weight at high temperature—that make them potentially superior to aluminum as the primary material for aircraft structures.

Despite their desirable properties, composite structures present difficult manufacturing and design challenges. One of the primary barriers to increased use of composites in commercial transports is manufacturing cost.[9] Currently, the carbon fiber-based thermoset composites that constitute the bulk of the composite materials used in commercial aircraft are too expensive to displace aluminum on a large scale. Yet despite the cost, airframe makers are convinced that the experience gained working with composites will bring costs down and constitutes a long-term investment in a critical capability.

There are two main areas of competitive activity in composites—fabricating structures and manufacturing basic materials. In fabrication, U.S. companies—including Boeing, McDonnell Douglas and others—have impressive capabilities on both the military and the commercial sides. The Japanese heavies possess superior capabilities in this area as well. They already supply composite structures such as tail cones and doors to both U.S. commercial airframe primes. In addition, MHI has developed through various programs culminating in the FS-X the capability to manufacture an entire composite wing in one piece through a process called "cocuring." The Japanese heavies have invested extensively in superior equipment (five-axis lay-up machines and autoclaves) for fabricating composite structures, and several companies have impressive R&D programs attacking key composites manufacturing issues. This invest-

---

[9]From the standpoint of an airframe manufacturer, the calculation is primarily one of price and performance. Testing new materials to ensure durability over the life of the aircraft is time consuming and expensive, but if the material performs and saves weight, and if its manufacturing costs do not raise its price, the airframe manufacterer will generally bear this expense.

ment and R&D activity indicate the importance that the Japanese aircraft industry places on developing world-class composites capabilities.

In the manufacture of basic composite materials—particularly carbon fiber—the Japanese position is even stronger than in fabrication. The current U.S.-Japan technological and competitive position in this area illustrates a number of the challenges the United States faces as DOD requirements become less important for driving the development and application of a range of technologies, particularly those relevant to the aircraft industry.

In the United States, DOD and the National Aeronautics and Space Administration (NASA) have provided major support over the years to develop a range of new advanced materials, and U.S. basic research at universities, and at national and industrial laboratories, is unmatched. Composites using carbon fiber have come furthest in their applicability, and a number of companies increased their production capacity in the late 1980s in anticipation of a large DOD demand base. However, since the late 1980s, the anticipated defense market has not materialized and a number of the U.S. manufacturers of carbon fiber have shut down or been sold to foreign investors.

Some of the leading producers of carbon fiber in the world are Japanese companies such as Toray, Toho, and Mitsubishi Rayon, which began making the materials to incorporate into sporting goods and other consumer products. This large manufacturing base has allowed them to focus on competing in the aircraft market with a longer-term view on the basis of competitive manufacturing costs. In addition to Toray's success in becoming the sole qualified supplier of carbon fiber and resin for the Boeing 777 composite tail, it has recently purchased the leading European manufacturer of carbon fiber. Toray did license a U.S. firm with its carbon fiber technology several years ago, but this did not result in establishing a price-competitive U.S. capability. A new Toray facility to be built near Seattle will weave and shape fibers made in Japan to Boeing specifications. Toray is interested in other aerospace applications, and in 1992 it purchased Composite Horizons, a small spin-off of Hughes Aircraft that manufactures composites for satellites.

Toray's competitive strategy and the nature of its alliances with U.S. companies highlight concerns about reciprocal technology transfer and market access in the field of advanced materials. For example, Toray has free access to the U.S. market, and is not restricted from working closely with Boeing and other lead users to hone its capabilities. It is also free to make manufacturing investments and high-technology acquisitions such as Composite Horizons. However, the committee heard that some U.S. materials makers have found it difficult to enter the Japanese market without forming an alliance with a Japanese company, often a potential competitor (although it is not a legal requirement). Such joint ventures generally do not provide opportunities for the U.S. partners to establish direct interactions with sophisticated customers in Japan who drive future development of components. The situation is evolving as U.S.

companies develop a variety of mechanisms to access growing markets for advanced materials in Japan and elsewhere.[10]

In contrast to the excellent but fragmented efforts of the United States in advanced materials, the Japanese approach of industry-government collaboration in this field leverages Japanese industry's existing strength in mass produced materials and incorporates focused government-funded research programs to target emerging applications. Basic research is conducted at a much lower level than in the United States, while basic research in U.S. universities is readily accessible to Japanese companies.[11] Government-industry technology development programs tend to focus on processes that optimize the utility of existing fibers and materials that are widely available. Aircraft structures and propulsion are major applications targeted in these programs. Commercial and military-oriented investments are mutually supportive.

It is clear that the Japanese government and Japanese industry see materials development as an important entry point to participation in future international aircraft programs. In order for the United States to reap the economic rewards of the substantial R&D funds expended in this area, both government and industry need to face up to several new challenges. For government, funding R&D on materials that must "buy their way onto the airplane" will require different criteria and research mechanisms than the "performance at any price" imperatives of military-driven technology development. For U.S. industry, it will be necessary to build better collaboration between materials suppliers and users than has been exhibited on the commercial side up to now. In addition, the challenge facing U.S. makers of advanced materials in accessing the Japanese market remains considerable.

While Japan's advanced materials capability has progressed to the point where Toray is supplying the material for the largest composite primary structure to date made by the U.S. aircraft industry, cuts in defense demand have led to severe distress for U.S. manufacturers of carbon fiber, causing several to exit the business.

---

[10]Du Pont has opened a laboratory in Europe to gain access to advanced materials users there, and has formed an alliance to improve access to Japan and Asia-Pacific markets. See Michael Mecham, "Du Pont Seeks Partners at Euro-Composite Center," *Aviation Week and Space Technology*, October 26, 1992, pp. 64-65; and "Du Pont, Mitsui Form Asia Region Composites Alliance," *Chemical and Engineering News*, December 13, 1993, p. 12.

[11]According to a 1991 report by an expert panel examining Japan's composites technology under the auspices of the Japan Technology Evaluation Center (JTEC), there is "little research effort in the fundamentals which determine the materials system selection or in the fundamentals of composite behavior. The Japanese were familiar with the systems selected for development in the United States and the rest of the world." R. Judd Diefendorf, Salvatore J. Grisaffe, William B. Hillig, John H. Perepezko, R. Byron Pipes, and James E. Sheehan, *JTEC Panel Report on Advanced Composites in Japan* (Baltimore, Md.: Loyola College, 1991), p. 13.

49

ENGINES

Because jet propulsion is the key enabling technology underlying commercial and military aviation as we know it today, the engine industry plays a special role in the aircraft supplier base. Both U.S. engine primes—GE Aircraft Engines and the Pratt & Whitney division of United Technologies—have extensive, long-standing technology linkages with Japan. The global context is important. Figure 3-3 illustrates the complex web of current international alliances in the commercial and military jet engine businesses. Both companies have been involved with Japan in military, commercial, and Japanese government-sponsored R&D programs.

GE has focused its engine collaboration in Japan with IHI,[12] while IHI—as the leading Japanese company in aeroengines—collaborates with Pratt & Whitney and Rolls Royce as well as GE. GE-IHI linkages have a longer history on the military side. GE was involved with the first Japanese postwar military aircraft program starting in 1953, with the J47 engine for the Japanese version of the F-86 fighter. Over the next several decades, GE's J79 engine was chosen to power the Japanese versions of the F-104 and F-4. GE's relationships with Japan during this period involved sending kits to IHI for assembly and test, with some components manufactured by IHI. More recently, GE's F110 engine was selected as the engine for the FS-X, and IHI is collaborating with GE in developing interfaces for the aircraft.

GE's collaboration with IHI in the development of a large commercial engine is fairly recent, having only begun with the GE90. The GE90 is the first of what GE hopes to be a new family of large engines to power the next generation of commercial transports. When the program was conceived in the late 1980s, it was decided that a global program structured around GE's existing international relationships would best leverage resources. In addition to IHI, which has an 8 percent share in the program, Snecma holds a 25 percent share and Fiat 8 percent. Each partner is responsible for designing and developing its specific part of the engine. IHI is responsible for several stages of turbine disks for the low-pressure turbine, the blades in those disks, and the long shaft that goes between the low-pressure turbine and the fan. Further, program participation requires partners to make considerable capital investments in testing and manufacturing infrastructure. IHI has proceeded to make the necessary investment to build a test cell.

The GE90 is currently undergoing testing and certification, and is scheduled to enter service in 1995. Although it is not possible to assess the bottom-line impacts on the participants, GE is pleased with the partnership and with IHI's contribution and performance to this point. The disks and turbine blades were impeccably designed and manufactured the first time around. GE has also learned some useful lessons from IHI, particularly from the rapid prototyping that IHI did for the turbine blade casting.

[12]GE's relationships with IHI and Toshiba date back to the pre-World War II period.

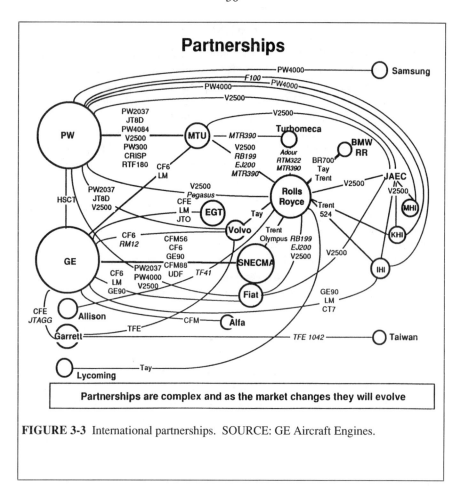

FIGURE 3-3 International partnerships. SOURCE: GE Aircraft Engines.

GE and IHI collaborate in several other areas. The HYPR program is discussed below. In addition, in July 1992 the two companies signed a broad MOU to jointly develop selected technologies. GE initiated the MOU because it realized that opportunities to learn from IHI will increasingly arise as IHI develops its own technologies through independent efforts and as part of Japanese government-sponsored programs. GE would provide some of its know-how in exchange. The MOU provides an umbrella structure for identifying and pursuing specific opportunities.

GE's formal technology transfer procedures are followed on each specific program undertaken with IHI (or any other partner). First, the business unit that wishes to transfer technology applies to a senior management technology council, which approves or disapproves specific transfers in light of the overall strategic position of GE Aircraft Engines. If the technology transfer is approved at this level, GE then submits an application to the Department of State for an export license, and to DOD and Department of Commerce as necessary. GE's licensed production contracts with IHI—going back to the J47—include flow-back provisions in which GE will obtain improvements that IHI makes in its technology.

Pratt & Whitney's (P&W) technology linkages with Japan are also extensive, and have included a slightly wider range of mechanisms and partners than GE's. P&W established a relationship with MHI in the 1930s that was interrupted by World War II, and it has also linked with IHI and KHI. P&W's motivations for establishing technology linkages with Japan are similar to GE's—to gain market access in military engines, to gain access to high-quality components, to spread development burdens, and—increasingly—to gain access to Japan's growing technological capabilities.

In 1978, the F100 engine was selected to be used on Japan's F-15s. This relationship has evolved from complete engines delivered to IHI, to knock-down kits, to licensed production. Some of the materials and the electronic engine controls were held back by DOD, but IHI now manufactures about 75 percent of the engine by dollar value. IHI continues to incorporate improvements that P&W developed for the U.S. version of the F100. P&W launched an earlier and less extensive military licensed production agreement in 1971 with MHI covering the JT8D-9 engine for Japan's C-1 military transport. In 1984, MHI became a 2.8 percent risk-sharing partner in the manufacture of a derivative product, the 20,000-pound JT8D-200.

P&W has two Japanese partners in the PW4000 program, a large engine with several derivatives that powers some versions of the Boeing 747, 767, 777, and the Airbus A300, A310, and A330 aircraft. The engine was originally developed in the early 1980s. Kawasaki became a 1 percent risk-sharing partner in 1985, and has continued at that level since then. MHI signed on as a 1 percent risk-sharing partner in the PW4000 program in 1989, and its participation grew to 5 percent in 1991 and 10 percent in 1993. MHI is responsible for manufacturing various turbine blades and vanes, turbine and compressor disks, active clearance control components, and combustion chambers. The increase in MHI's share since 1989 has come about as a result of mutual satisfaction with the relationship and a desire to expand it.

In addition to risk-sharing agreements with MHI and KHI in commercial engines, P&W has a long-term sourcing agreement with IHI to produce the long shaft connecting the high- and low-pressure turbines for the JT9D, PW2000, and PW4000. IHI will manufacture all of Pratt & Whitney's commercial long shafts. Utilizing and improving upon the process transferred in connection with the F100 program, IHI has become a world-class center for the production of long shafts of more than 8 feet. As mentioned earlier, IHI will be manufacturing the long shaft for the GE90, and it manufactures all of Rolls Royce's shafts as well. This specialization is not uncommon in the engine business: Fiat dominates the manufacture of gear boxes, and Volvo is strong in casings. Although IHI's dominance in shafts raises issues of dependence and possible supply disruption, the engine primes manage this dependence by maintaining some capability of their own. The focused manufacturing approach carries significant benefits in terms of cost and quality.

Pratt & Whitney is also a major partner in International Aero Engines (IAE), a global program that developed and is now marketing the V2500 engine. This program marked the first time the Japanese participated in a major engine development program. Pratt & Whitney and Rolls Royce are the lead partners—both hold 30 percent shares in the program. Germany's MTU holds 11 percent and Fiat 6 percent. Japan Aero Engine Company (JAEC) holds 23 percent of IAE, and is itself a joint venture of IHI (with 60 percent of JAEC), Kawasaki (25 percent) and MHI (15 percent). JAEC is responsible for the fan and the low-pressure compressor.

All of the non-U.S. members of IAE received support from their governments for their participation. JAEC has received annual payments of $20 million to $25 million from the Ministry of International Trade and Industry (MITI) since the start of the abortive FJR710 program in the early 1970s, and this support has continued through V2500 development, covering roughly 75 percent of JAEC's development costs, 66 percent of testing costs, and 50 percent of the production tooling and nonrecurring startup costs.[1] Repayment with interest of these success-conditional loans is slated to commence when the program breaks even. The V2500 faces tough competition from the CFM International CFM56, but appears to be gaining greater market acceptance over time.

Both GE and P&W participate in the Japanese Supersonic/Hypersonic Propulsion Technology Program (JSPTP or HYPR), which was launched by MITI in 1989 as a $200 million, 8-year program (since extended to 10 years). The ultimate goal of the program is the development of a scale prototype turboramjet, Mach 5 methane-fueled engine. The program is administered by MITI through its Agency of Industrial Science and Technology and the quasi-governmental New Energy and Industrial Technology Development Organization.

The Japanese partners—IHI, KHI, and MHI—receive 75 percent of the funding and take the lead on technology development and design. HYPR is significant in that it is one of the first of Japan's national R&D projects to contemplate international participation from the outset as an integral feature of the program. The foreign participants—who receive 25 percent of the funding—are Pratt & Whitney, GE, Rolls Royce, and Snecma. The formal agreement between MITI and the foreign engine companies was signed in early 1991. The process of negotiating this participation was somewhat long and complex, the major stumbling block being the treatment of intellectual property generated in the project. The four foreign companies joined together to negotiate with MITI as a united front. This process led to an agreement and a change in Japan's laws governing intellectual property rights in government-sponsored R&D.

---

[1]David C. Mowery, *Alliance Politics and Economics: Multinational Joint Ventures in Commercial Aircraft* (Cambridge, Mass.: Ballinger Publishing Company, 1987), p. 93.

From the point of view of GE and Pratt & Whitney, the main motivation for participating is that taking a role in the Japanese program is preferable to a major supersonic/hypersonic engine program going forward without U.S. involvement. By participating, GE and Pratt & Whitney gain insights into the basic design decisions and capabilities of the Japanese members of HYPR. Thanks to MITI funding, participation is not costly for the foreign firms. The U.S. engine makers believe that as a major terminus for flights of the next-generation supersonic transport, Japan will inevitably be involved in its development. As a separate initiative, GE and P&W are collaborating on NASA-funded research on high-speed civil transport propulsion targeting an engine in the Mach 2-2.5 range.

The basic interaction between foreign and Japanese companies in HYPR is participation in design review and analysis in designated program areas. Since the program is currently in its fourth year and will probably run for ten, the impacts and implications cannot be assessed precisely. The eventual impact will depend a great deal on the timing and mechanism for developing propulsion for the next-generation supersonic transport. While foreign participation allows the major international players to keep tabs on Japan's approach, the Japanese participants gain design insights from foreign coaching. Also, international participation in HYPR has itself served to give credibility to Japanese efforts to play a significant role in international advanced engine programs and to other Japanese government efforts to organize international R&D collaboration.

The Japanese government also funds several other programs that have implications for future aircraft propulsion systems. The one that is most closely linked to HYPR is the research program on high-performance materials organized under MITI's "Jisedai" or Next-Generation Technology Development funding pool. The program began in 1989 and is scheduled to run through 1996. In addition to these ongoing R&D programs, the Japanese government— mainly MITI and the Technology Research and Development Institute—is conducting a number of feasibility studies aimed at significantly upgrading Japan's engine testing facilities over the next decade. JDA is also making funds available for a high-altitude test facility in Hokkaido.

Japanese aircraft engine makers have effectively leveraged private and public resources in international alliances and public R&D projects to improve and deepen their technological and manufacturing capabilities. Individually or as a group, Japanese companies are well positioned to continue to participate in international engine development programs at increased levels of technical and manufacturing responsibility. Japan's government technology programs and corporate strategies are aimed at playing a major role, if not one of world leadership, in advanced propulsion materials and other targeted critical technologies.

AVIONICS

Avionics is another critical part of modern transport aircraft. Advances in navigation and flight control systems have the potential to further reduce the cost and increase the safety of air travel. Commercial avionics is a $3 billion per year business worldwide. The two dominant players are American companies—the Collins division of Rockwell International and Honeywell. U.S.-Japan technology linkages are fairly extensive in this sector and take several characteristic forms depending on the market.

On the commercial side, U.S.-Japan linkages have been driven by changes in the nature of innovation in avionics hardware over the past 15 years. Up to the mid- or late 1970s, the bulk of hardware innovations incorporated into commercial avionics came from military electronics developments. Increasingly, however, avionics systems incorporate component technologies first developed for consumer electronics and high-demand computer applications. Over the past several decades, as Japanese companies achieved and extended their dominance in consumer electronics and gained strong positions in several areas of the semiconductor industry, Japan has become the major source for these hardware innovations. Although standard components can be incorporated into avionics black boxes in some areas, in others the performance requirements necessary for an avionics application go so far beyond the capabilities of the standard component that an extensive modification effort is necessary. This is the fundamental dynamic driving U.S.-Japan technology linkages in commercial avionics today.

The best current example of this trend is flat panel displays. The liquid crystal display (LCD) technology that was invented in the United States in the late 1960s has been nurtured and improved by a number of Japanese companies for more than 20 years. Passive and active matrix LCDs are now the dominant technology of flat panel displays in rapidly growing markets such as portable computers and hand-held television sets. Japanese companies such as Sharp and Hosiden are the leaders in this technology, and Japan currently holds more than 90 percent of the flat panel display market. Several small U.S. firms develop and manufacture some displays for military and other niche applications, but they do not have the capital to invest in the necessary manufacturing capability for large-scale production.

In developing the next-generation avionics systems that will be installed in the Boeing 777, both Collins and Honeywell clearly saw the advantages—mainly space and weight savings—of replacing cathode-ray tube displays with flat panels. Although both Collins and Honeywell briefly considered other alternatives, it soon became clear that the Japanese companies that currently dominate the world market were the best source of a cost-efficient solution. Collins teamed with Sharp and Toshiba, and Honeywell worked with Hosiden.

Whereas the necessity for acquiring this high value-added component was clear and compelling on the U.S. side, the Japanese display makers had to be convinced to take up the task—avionics is not a large market compared to laptop computers, and a significant commitment of engineering resources would be required. However, there were also compelling advantages for the Japanese display makers, such as the opportunity to lock in a long-term, profitable stream of business and to develop new capabilities for their displays.

Perhaps the most important benefits for the Japanese firms were the interrelated benefits of learning about technology and business methodology in a very high-image market. Although the American firms were very careful to employ the Boeing-like strategy of keeping these key suppliers limited to display development, technology transfer was necessary to enable the Japanese companies to solve the unique problems arising in the development of displays that meet avionics needs.

In terms of the immediate business and technical objectives, U.S.-Japan linkages in commercial avionics usually achieve their goals and bring the expected benefits to both sides. The U.S. integrator gains a reliable supply of high value-added components or subassemblies at a reasonable price, which helps add value for the end user. The Japanese partner gains steady business, technology, and learning benefits that can be applied to its core business or serve as a basis for further expansion in aircraft markets. For example, many new aircraft will incorporate flat panel displays in the cabin as part of passenger entertainment and communications systems as well as in avionics. The Japanese display makers can directly apply knowledge of the business methodologies of airframe makers and airlines to their efforts to market displays for these systems.

The downside of the flat panel display relationship was felt when the small American manufacturers filed an antidumping suit against the Japanese, and the International Trade Commission placed punitive tariffs on Japanese imports. Collins and Honeywell have been hurt by these duties, but would not consider transferring manufacturing out of the United States in response, as have several U.S. makers of laptop computers.

There are also extensive U.S.-Japan technology linkages in military avionics, but these are of a completely different character from the commercial supplier alliances. In order to gain access to the JDA market, U.S. avionics makers must often license production or enter other collaborative relationships with Japanese companies like Mitsubishi Electric or Japan Aviation Electronics. This often happens as part of a licensed production program such as the F-15.

In all of these relationships, the transfer of technology is almost exclusively from the United States to Japan. In the commercial field, the United States receives products in return for technology; and on the military side, market access.

What are the competitive issues posed by U.S.-Japan linkages in avionics? Is there a long-term danger that Japanese companies will become full-line avionics integrators? American avionics industry leaders recognize the considerable technological and manufacturing capabilities of Japanese electronics companies, and realize that they often deal from a position of weakness in seeking to gain access to component technologies. A high percentage of the value added in current avionics systems consists of Japanese components, and the percentage will very likely continue to rise. The avionics market might be appealing for the Japanese as a high-profile industrial market with some potential for driving technology development that could be applied to core businesses. Also, the Japanese are very aggressive in developing nonavionics electronic systems for aircraft (entertainment systems and satellite communications), as well as mass market applications of technologies that are closely related to avionics, such as automotive applications of GPS (already being marketed in Japan). Collins and Honeywell are interested in some of these markets, but they are not well entrenched.

However, while Japanese companies are capable of moving up the avionics food chain, there are significant capabilities that they do not yet possess, and there are few signs that Japanese industry or government is aggressively developing them. For example, the software and systems integration skills that are needed to develop the current generation of avionics is beyond current Japanese experience. Although the value of Japanese components is high, U.S. avionics companies do not anticipate a short-term challenge from Japan in the integration segment. They would prefer to have more leverage as they incorporate Japanese technologies into their systems, but believe that the lack of a U.S. consumer electronics industry is the main cause of the difficulties they have experienced.

From a U.S. policy perspective, the impacts and implications are more complex. The U.S. Department of Defense and other agencies have identified flat panel displays as a key technology for a range of industries. DOD's Advanced Research Projects Agency, the Department of Commerce's Advanced Technology Program, and Department of Energy laboratories have launched a number of technology development programs to help build competitive U.S. capabilities in this area. Avionics companies might be obvious "lead users" in these efforts, but it does not appear that any U.S. avionics companies are currently involved in U.S. government-sponsored efforts. As in advanced materials, the flat panel display example illustrates that the challenges involved in planning and implementing an effective civilian technology policy are considerable.

OTHER COMPONENTS AND SUBSYSTEMS

Modern transport aircraft incorporate a large number of subsystems and components manufactured by a variety of large and small companies. Subsystems include electrical power systems, actuation systems, and landing gear. Each subsystem and the aircraft as a whole incorporate numerous and varied components such as gears, materials, and integrated circuits. Japanese companies have become quite prominent in some parts of the supplier base. For example, most of the precision bearings needed for aircraft engines are now manufactured in either Japan or Germany. Companies in those countries built on existing strengths in bearings to eliminate the remaining American companies from this high-performance segment. Bearings—like flat panel displays—represent a field in which Japanese companies entered aerospace markets because of capability acquired in more general-purpose markets.

Japanese success in supplying dedicated aircraft components and subsystems is more uneven. Teijin Seiki—whose original business was textile equipment—has achieved a prominent position in primary actuation systems and supplies all recent Boeing programs. Teijin Seiki is also actively building other parts of its aircraft subsystem business, partly through a joint venture with Sundstrand. Other Japanese companies such as Kayaba, Shinko Electric, and Yokohama Rubber have also gained success in some subsystem and component areas.

Because of the wide variety of products and companies involved, U.S.-Japan technology linkages in aircraft subsystems and components are difficult to characterize in a general way. In contrast to expanding relationships between U.S. primes and Japanese suppliers, there appear to be few linkages between U.S. and Japanese suppliers on the commercial side. Most U.S. supplier-Japanese supplier technology linkages have been formed in the context of Japanese military programs. Particularly in cases where U.S. systems have been coproduced or produced under license in Japan, JDA and Japanese industry generally pursue licensed production of U.S. subsystems and components that embody significant technology.

The STS Corporation joint venture between Sundstrand and Teijin Seiki raises a number of the relevant technology transfer and market access issues faced by U.S. suppliers wishing to participate in the Japanese aircraft market. Sundstrand's involvement in the Japanese aircraft market began in the late 1960s with licensed production of electric power generating system constant-speed drives by Teijin Seiki for Japanese military programs. This licensing arrangement evolved into the formation of a 50-50 joint venture company called STS Corporation about a decade ago.

Improved market access was Sundstrand's primary motivation, and it has seen tangible benefits in this regard. The original target was the military market, and a significant proportion of STS's sales still go to military programs. There appear to be market access benefits on the commercial side as well. Examples include STS's supply of the main fuel pump for the V2500 engine and Sundstrand's participation in the MD-12 actuation team with Teijin Seiki and Parker Hannifin.

As the venture markets Sundstrand's more mature technologies, the transfer of technology through the joint venture has been predominantly from Sundstrand to STS—while Teijin Seiki provides the personnel to staff the venture. STS participation in SJAC collaborative R&D programs may bring reverse technology transfer opportunities in the future.

Although the venture should be termed a success from the standpoint of the strategies of the two parent companies, it is an illustration of continuing U.S.-Japan technology and market access asymmetries. While Sundstrand found it prudent to team with a Japanese company in order to expand market opportunities in Japan (and even in the United States, in the case of the MD-12), there are few barriers to Teijin Seiki and other Japanese subcontractors selling directly to Boeing and other U.S. primes.

DISTINCTIVE FEATURES OF U.S.-JAPAN AIRCRAFT LINKAGES

Motivations and Benefits for the United States

The committee identified a number of significant motivating factors and benefits of expanding U.S.-Japan technology linkages in the aircraft industry. Significantly, these benefits are more likely to be realized by U.S. companies dealing from the strongest technological and business positions—the airframe and engine primes such as Boeing, Pratt & Whitney, and GE. Most of these benefits revolve around the generally superior performance of Japanese companies in this field as business partners.

Long-Term Commitment of Resources

Japan as a country, as well as the individual companies involved in the industry, is committed to a long-range strategy of aerospace growth. U.S. companies linking with Japan can be reasonably assured of their partner's commitment to invest despite the long-term payoffs typical in this industry. This is a particularly important benefit in view of the escalating costs of aircraft and engine development programs.

No Barriers

Apart from the difficulties that U.S. companies may experience in acquiring Japanese companies, there appear to be no formal barriers to aerospace cooperation—at least at the level of U.S. prime integrators.

Focus

Japanese partners generally have an unrelenting focus on meeting project schedules and target costs, once negotiated. They bring their notions of continuous improvement to the program and do not hesitate to invest to constantly improve quality.

Access to World-Class Manufacturing

Japan's focus on manufacturing processes is apparent throughout their aerospace units. Much of this know-how is available to U.S. partner companies if they are willing to make the investment needed to transfer the technology back home.

No Leakage

In the aerospace industry today it is common for individual Japanese companies to have partnerships or close relationships with several competitive companies at the same time. This provides an opportunity for "leakage" of plans, activities, and know-how from one competitor to another. U.S. companies who have partnered with Japanese aircraft companies have not experienced this problem.

U.S. Access to the Japanese Market

Although Japan has no aircraft offset requirements or other formal trade barriers in aircraft, market leverage has been a major motivator of linkages on the U.S. side. Airbus has managed to sell aircraft to Japan despite an absence of significant linkages. Generally speaking, however, a presence in Japan is seen as a prerequisite for participation in the market. Boeing and McDonnell Douglas have sourced portions of airframes in Japan to enable or promote sales to the Japanese airlines. Engine makers have also followed this strategy, whereas components manufacturers have followed a pattern of joint ventures as a means of gaining access as suppliers to the market for military aircraft in Japan. U.S. companies have also found, however, that cooperative programs do not ensure sales.

Risks for U.S. Industry and the United States

Juxtaposed to the benefits are the risks of technological collaboration with Japan. The committee identified a number of risks faced by participants in linkages and other U.S. companies.

Enabling Competitors

One risk faced by U.S. companies undertaking technology alliances with Japanese companies (or other foreign partners) is that technology transferred to a partner through the alliance, or developed independently thanks to the joint program revenue base, will be used by the Japanese partner to market a competing product. This can occur through military or commercial programs. At the prime airframe and engine levels, this risk has been realized by U.S. firms, but not in relationship to Japan so far—some technologies, such as fly-by-wire transferred through the European F-16 coproduction program, were later used by Airbus. Some U.S. suppliers, particularly in the context of military programs, have faced Japanese competition from licensing partners—the ring laser gyro case cited in Appendix B is one example.

Displacement

Another risk faced by U.S. suppliers is that they will be displaced by Japanese competitors in the context of technology linkages with Japan formed by other U.S. companies. This is currently occurring in the aircraft structures area, for example, the displacement of Northrop as a components supplier on the 767 and 777.[2]

Dependence and Loss of Critical Capabilities

In some cases, Japanese strength in technologies developed in other industries and applied to aircraft may have the effect of stifling nascent U.S. capabilities. To U.S. industry, there is a danger that capabilities critical to the future of the industry will be completely absent in the United States, or that U.S. basic research feeds development and commercialization activity that largely occurs in Japan (as has happened in industries such as robotics). If the technology is critical enough, U.S. primes may find themselves in the position of having to transfer more of their own technology than they would like in order to access Japanese capabilities. Flat panel displays and some areas of advanced materials are examples in which this risk has been realized.

Market Access Problems

At both the prime and the supplier levels, but particularly in the case of suppliers, formal and informal Japanese trade and investment barriers necessitate the trade of technology for market access. Often, the only viable option that makes business sense to U.S. companies is a joint venture. In some cases this may be a low-risk strategy. In areas where direct contact with customers plays a major role in driving technology development, however, some companies have found that their joint ventures constrain their relations with other Japanese companies or serve to create powerful competitors.

Technology Access Problems

U.S. companies and the United States as a country have transferred far more technology to Japan than vice versa, particularly in the aircraft industry.

---

[2]Northrop had the opportunity to compete for supply of components for the 777.

The United States runs the risk of forgoing significant opportunities to improve the competitiveness of its aircraft industry if the flow of technology from Japan is not increased—both through lowering Japanese barriers and through devoting more of its own resources to acquiring and assimilating the available information.

National Security

Particularly in military programs, there is always a risk that technology transferred overseas could come back to threaten U.S. national security; this is the rationale behind export controls. In the case of Japan, this risk has been judged to be quite low—a qualitatively higher level of technology transfer through aircraft and other military programs has been allowed for Japan than for other allies.[3]

Evolution of Linkage Mechanisms

U.S.-Japan linkages display several characteristic features in terms of mechanisms and trends:

• For linkages formed by U.S. integrators in airframes and engines on the commercial side, interaction with Japanese companies generally begins with the establishment of a supply relationship. In the case of linkages of U.S. primes with the Japanese heavies, interaction has increased over time—often supported by Japanese government policy. Japanese companies are now or could ultimately become capable of manufacturing all the parts of modern commercial airframes and engines. In engines and airframes, the pattern may be shifting toward significant Japanese participation as partners in global programs managed by U.S. primes.

• A relatively new mechanism is foreign participation in Japanese government-sponsored R&D programs such as HYPR. The future direction of this mechanism is unclear, but in addition to true joint development of new technologies, such programs also have a component of transferring existing knowledge from foreign firms with advanced capabilities to their Japanese competitors. Although the Japanese government has launched several international R&D programs since HYPR, none has focused on aircraft specifically.

• At the prime level in military programs, the pattern has been one of U.S. technology transfer to Japan, with a dynamic of increasing Japanese responsibility and technological capability over time. The direction of U.S.-Japan military program links after the FS-X is unclear.

• Linkages at the supplier level present a mixed picture. On the military side, U.S.-Japanese supplier links often occur in conjunction with large pro-

---

[3]Japan Aviation Electronics, however, violated export controls and was sanctioned by the government of Japan for shipping ring laser gyros to Iran.

grams and involve Japanese licensed production of the U.S. component. In cases where the U.S. company has a strong technological edge, these relationships sometimes extend over several programs and even evolve into collaboration in commercial fields. In other cases, such as the mission computer and the electric power generating system for the FS-X, Japanese companies have displaced U.S. companies by developing independent capabilities.

U.S. and Japanese Strengths and Weaknesses Underlying Linkages

Japanese Strength—Manufacturing Capability and Investment Resources

Japanese aircraft companies have demonstrated the creativity and resource commitment necessary to apply world-class technology to aircraft production. In addition to the aircraft industry itself, Japanese capabilities in areas such as composite materials and flat panel displays have been developed through investment in manufacturing excellence aimed at other markets, and are finding increasing application in aircraft. In contrast, some U.S. companies have found it difficult to invest in capital-intensive manufacturing processes in the United States in recent years.

Japanese Strength—Integrated, Supportive Policy Environment

A Japanese policy environment encouraging international alliances that transfer technology to Japan, civil-military integration in the domestic industry, and cooperation between companies helps to maximize the impact of Japan's technological strengths. Although the U.S. aircraft industry is dynamic, policy agendas are often fragmented and government agencies sometimes work at cross-purposes.

U.S. Strength—Systems Integration and Other Advanced Technologies

U.S. technological excellence across a wide range of aircraft technologies—particularly those associated with systems integration—is unmatched. U.S. industry, academic, and government R&D capabilities in aeronautics, propulsion, materials, and other associated fields are the foundation for future U.S. competitiveness in the global aircraft industry. Although Japan is making efforts to build wind tunnels and other necessary research infrastructure, considerable resources over a long time period will be necessary.

U.S. Strength—Long-Term Familiarity with Needs of the Global Market

Particularly at the level of integrating airframes, engines, and avionics, U.S. companies have maintained an aggressive global marketing presence that facilitates the incorporation of customer needs into products, as well as the capabilities in safety certification necessary to sell products globally. The Japanese industry has tried to develop marketing and product support capabilities though international alliances, with some limited success.

Outcomes and Implications of U.S.-Japan Linkages

Japanese Capabilities and Strategy

Although Japan is missing some technological squares in the matrix of critical aircraft capabilities (systems integration, marketing), it currently possesses the necessary infrastructure to support an indigenous aircraft industry. Japanese companies and government are pursuing international alliances and technology development programs to fill in the missing pieces. Japan is making the necessary investments to increase its presence in the commercial aircraft market, focusing on manufacturing quality and cost leadership.

Japan has not launched an effort at the airframe or engine prime level to compete with U.S. firms; nor has it formed significant relationships with Airbus or other international players. Japan would likely become a formidable U.S. competitor if it decided to pursue either of these options, and government and industry are currently reevaluating their basic approach to the industry. In any case, Japan is a significant factor in the global aircraft industry. U.S. collaboration with Japan entails benefits and also some risks, but at this point it appears that continued cooperation is preferable to the alternatives.

Technology Transfer

Technology flow through U.S.-Japan linkages in the aircraft industry has been predominantly from the United States to Japan. Although historically, more technology has flowed through military than commercial programs, commercial alliances formed over the past 10 to 15 years have also transferred technology to Japan or in some cases, stimulated independent Japanese development when they were not given access. Although it appears that DOD and U.S. companies involved in military and commercial linkages have by and large protected critical technologies while reaping significant benefits from these relationships, the impacts of the most recent and significant technology transfers (through the FS-X and extensive commercial program links) are still unclear. The security environment that justified a pattern of extensive U.S. aircraft technology transfer to Japan is rapidly changing, and there is a need to take economic considerations into account. In view of the large U.S. stakes in this industry and the rapidly expanding Japanese capabilities in many significant technologies, a more balanced flow of aircraft technology between the two countries should be key to a continuation of mutually beneficial interaction, and should be pursued by U.S. industry and government as a strategy and a major policy goal.

The U.S. Supplier Base

Although U.S. primes have by and large had good experiences in their relationships with Japan, evolving patterns of global manufacturing capability and industry restructuring—in which U.S.-Japan linkages are an important part of the context—already threaten existing parts of the U.S. supplier base and may prevent the development of U.S. commercial capabilities in a number of critical, emerging areas. This situation suggests the need to reexamine the manner in which technology development and related business activities are organized and funded in the United States, in order to promote more effective relationships between U.S. companies and between industry and government, as well as ensure retention of an innovative full-spectrum aerospace capability. Where the technologies impart both security and economic growth, there is a need for more attention and coordination among various government agencies to ensure effective use of public support for R&D and procurement relevant to industry, especially the supplier companies. Failure to address these issues implies continued erosion of the domestic U.S. supplier base and a concomitant increase in the probability of Japanese entry at the prime level as its supplier base becomes more developed.

4

Future Trends

MARKETS

Over the next several decades, demand for air transport, and the aircraft necessary to carry it, should continue to grow at a relatively high rate. Boeing's forecast, for example, envisions an annual average growth in world airline passenger traffic of 5.4 percent from 1992 through 2010.[1] Such rates are somewhat lower than those prevailing over the past two decades (6.8 percent), but are still substantial. Similar trends are expected for the global air-freight market (with a 6.5 percent growth to 2010 in the Boeing forecast, compared to 8.0 percent from 1970 to 1992).

This overall growth will produce a continued increase in demand for new commercial aircraft. Boeing anticipates a global market for aircraft of $815 billion in constant 1992 dollars from 1992 to 2010 (including $204 billion in replacements and $611 billion in additional capacity).[2] McDonnell Douglas envisions a somewhat higher $1.0 trillion global market for aircraft over the same time period, representing a total of 14,072 units.[3]

---

[1] Boeing Commercial Airplane Group, *Current Market Outlook: World Market Demand and Airplane Supply Requirements* (1993), p. 25.

[2] Boeing, ibid., p. 34.

[3] Estimates provided by McDonnell Douglas, May 1993.

Within this overall picture of growth of traffic and aircraft demand, important market shifts will take place, with the nations of the Asia-Pacific region experiencing higher economic and air traffic growth than other regions of the world. This region has had some of the most rapidly growing economies in the world, and most forecasts anticipate that this pattern will continue. Furthermore, integration of China, and now possibly Indochina, into the regional and global marketplace is continuing to lead to expanded international travel associated with it. Although the region faces some uncertainty on the security front in the post-Cold War era, even these potential problems seem less serious than in most other parts of the world. Boeing foresees that intra-Asian travel will grow at an annual rate of 8.4 percent to 2010, and transPacific travel at 7.8 percent, substantially above the global average. Some of the Asian and Asia-Pacific international traffic will move on U.S.-owned airlines, but the market shift will also involve an increased role for non-U.S. airlines. McDonnell Douglas, for example, anticipates that the dollar value of aircraft deliveries to the Asia-Pacific countries will account for 39 percent of the global total. Whereas only Japan and Australia are among the top 10 countries in terms of the dollar value of aircraft deliveries through 1992, Asia-Pacific countries will account for 5 of the top 10 from 1993 through 2010 in the Boeing forecast (with Japan and Australia joined by China, South Korea, and Singapore). Japan alone is anticipated to account for $60.5 billion of the overall market for new planes through 2010 (7.4 percent of global demand).

This long-term optimism about the market for new commercial aircraft contrasts with considerable pessimism concerning the next several years. Most observers believe that sales will continue to decline from their 1992 peak for several more years, perhaps through 1996. Airframe and engine makers are counting on a surge in new orders from U.S. airlines for delivery later in the decade (aided by requirements to meet more stringent noise regulations).

Although we can be reasonably certain that growth in the demand for air travel and environmental regulation will lead to more aircraft sales over the coming decade and beyond, continuing financial pressure and structural change in the U.S. and global airline businesses may permanently affect traditional purchasing criteria. Airline deregulation, both in the United States and internationally, implies a continuation of strong price competition among airlines leading to average profit levels over the next several decades that will remain lower than in the past. In this more stringent competitive environment, airlines will put increased pressure on the manufacturers to cut prices. This situation is exacerbated by the longevity of aircraft. Even with improved performance characteristics, airlines demand that the anticipated ownership and operating cost of the new aircraft represent a significant savings over existing aircraft. This represents a change from the past when higher profits gave the airlines more financial leeway to introduce new models.

Such a shift in airline demand will lead the aircraft industry toward increased price competition; the most successful participants in all aspects of the industry will be those companies that can reduce manufacturing costs in order to maintain profitability at lower prices. Even with cost reductions, however, price competition is also likely to result in lower profit levels within the industry relative to the past, which could affect the funds available for research and development on next-generation products.

The boom-bust cycle of the airline business since deregulation in the United States is a reminder that the market outlook can change very quickly. Faster than anticipated economic growth could reopen the financial spigots and contribute to partly rebuilding the long queues for aircraft that existed several years ago. Yet even under the most optimistic assumptions, competition is likely to be fierce in most segments of the large commercial transport market.

Trends in military demand will have an indirect but significant impact on the commercial industry. Because of the steep projected declines in U.S. aircraft procurement, overall U.S. aerospace industry restructuring is likely to further accelerate and build toward a climax over the next several years. Foreign demand for U.S. military aircraft has also declined recently, and there appear to be few signs that it will pick up enough to offset much of the U.S. procurement decline. Although the political factors that influence military aircraft demand are even more difficult to foresee than commercial trends, the important point is that restructuring strategies for diversification, acquisition, and divestment are now being formulated on the basis of the current outlook. Although military and commercial businesses (including manufacturing and design) are often separated in U.S. companies—more so in airframes and structures than in engines and some component areas—many of the significant aircraft defense contractors are players in at least some aspect of the commercial business. Therefore, military restructuring has the potential to spur significant shifts in the U.S. commercial transport business. Since a number of the commercial businesses are not as visible or sensitive as prime military work and could easily be spun off, even some global consolidation through foreign investment in given industry segments should not be ruled out.

Overall, this mixed scenario for short- and long-term market trends holds the danger that some firms with good long-term prospects, mostly at the components level, will fall by the wayside. For all firms at all levels of the industry that do survive the short-term problems, success will be increasingly dependent on an ability to cut costs while maintaining or enhancing quality.

NEW PROGRAMS

A number of new products are in the advanced stages of development and will enter service within the next few years.[4] At the airframe and engine prime levels, new programs are currently difficult to finance. The traditional methods

---

[4]In aircraft, the Boeing 777; the McDonnell Douglas MD-90; the Airbus A321, A340, and A330; and the Ilyushin IL-96M; in engines the General Electric GE90, the Pratt & Whitney PW4087, and the Rolls Royce Trent.

(advances and downpayments from airlines, and commercial adaptation of engines developed for the military) have become all but unavailable over the past decade. The resulting need to raise risk-sharing capital has been one of the driving forces behind the growth in international alliances in this industry.

Much attention is focused on the possible development of a new generation of very large transports that could carry from 400 to more than 600 or even 800 passengers. McDonnell Douglas's preliminary design for the MD-12 would be at the lower end of that capacity range, but the company is reluctant to launch the program without a major equity or risk-sharing partner. A planned partnership with Taiwan Aerospace fell through in 1992. Both Boeing and Airbus are conducting feasibility studies and holding preliminary discussions with potential partners concerning even larger airplanes. Japan is considered central to these discussions. Most experts believe that there is a market for one of these very large planes, and the bulk of that market is in either transpacific or domestic Japanese routes.[5] The relatively limited size of the global market, in terms of the total number of such planes that would be produced, leads industry experts to believe that efficiency in production implies only a single producer. Furthermore, the high development costs associated with a very large aircraft suggest that firms will have difficulty convincing capital markets or governments to supply the necessary capital. Both efficiency and capital access considerations point toward an international consortium. For example, as a result of the U.S.-European Community subsidy agreement of 1992, Airbus may no longer have easy access to member government funding for program launch. The consortium has been actively courting the Japanese "heavies."

In other range/passenger categories, it will likely be difficult to launch all-new programs even if business improves. Boeing has, for example, recently decided to develop an advanced version of its 737 rather than a completely new airplane in the 100 to 150-seat segment. Since this was the projected capacity of the 7J7-YXX, the Boeing decision appears to have dealt a blow to Japan's immediate prospect for taking a more significant partnership role in a new program. Apart from the 80 to 100-seat and the more than 400-seat categories, as well as the supersonic arena, which is discussed below, there are no obvious unaddressed market needs within the current market framework.

ADVANCED TECHNOLOGY

One focus of advanced technology development for aircraft is enabling technology for a second-generation supersonic transport. The major issues are environmental (noise and emissions), and the keys to resolving them are in the propulsion and propulsion-airframe integration areas. Even if significant progress is made on the technical front over the next several years, an actual High

---

[5] 747s are used much less on transatlantic than transpacific routes. Japan is the only country that uses 747s on domestic routes.

Speed Civil Transport (HSCT) development program will be very difficult to launch. Airframers will not commit to development unless they are certain that the environmental impact will be acceptable. Also, as in the case of a very large transport, the nearer-term projected market (to about 2015) appears to allow room for only one program (although the HSCT market is expected to grow considerably in the long term). Since the bulk of the market is global and transoceanic, most observers anticipate a global program of some sort, but structuring such a program will be a complicated and difficult undertaking. Finally, extensive international government involvement, in areas such as environmental and safety certification, and infrastructure (if not program financing), will likely be necessary. Although it is possible that an HSCT will be flying at some point during the first decade of the next century, substantial technical and business-related obstacles remain.[6]

Although maintaining leadership in HSCT-related technologies is critical for the U.S. aircraft industry in the long term, *a more immediate concern is the development of technologies that can be incorporated into advanced subsonic aircraft.* The major issues are those affecting cost and quality, including process and manufacturing technologies. Airlines will demand the superior performance made possible by new technology but will not be in a position to pay premium prices. Therefore, the incorporation of new technology must not only "pay for itself" in terms of lower operating costs over the life of the aircraft, but also avoid increasing the initial unit cost. Process technology developments with the potential for raising quality while lowering the unit cost of aircraft could also affect the competitive landscape during the coming decade.

IMPACT OF BROAD INDUSTRY FORCES

The committee has identified several broad trends for the aircraft industry during the coming decade—growing but price-sensitive markets, global restructuring, and few new programs launched by the established players. What do these trends and specific regional factors imply?

• Over the coming decade, one of the keys to survival and growth in the global aircraft market will be manufacturing performance in terms of low cost, high quality, and prompt delivery. Companies at the prime level through almost all parts of the supply chain will feel continuing pressure to achieve higher quality at lower cost.

• Japanese aircraft companies are currently investing heavily in manufacturing technology. Although a few U.S. companies are making the necessary long-term investments, many are not.[7] Japanese industry is likely to tighten its hold in current areas of excellence (structures, composite materials, engine

---

[6]In the event of significant continuing distress in the global aircraft industry, political pressure may mount in producing countries to launch an HSCT as an industrial policy measure.

[7]U.S. Census Bureau data showed a 25 percent drop in capital spending by aircraft manufacturers for the first half of 1993 compared to the previous year.

components, flat panel displays, other electronic components, and other component systems such as primary actuation).

- While consolidation of the Japanese position at various supplier levels will ensure that the trend toward increasing Japanese value added in commercial aircraft continues, competitive and financial pressures on U.S. primes and on suppliers outside the current scope of Japanese activities will also continue. It is now close to impossible to predict where this might lead and what opportunities (expanded partnerships or direct investment) might be available to U.S. and Japanese companies as a result.

- In addition to its advantages as a manufacturer, if Japanese industry can retain its traditional access to long-term capital, it will likely gain more leverage in partnership negotiations for new programs. The strong economic expansion, accompanied by extremely low interest rates and financial asset inflation, gave all Japanese manufacturing a temporary advantage in raising capital in the late 1980s. Although this so-called bubble is over in Japan, the aircraft industry remains high on the government agenda for industrial promotion, meaning that access to policy lending from the Japan Development Bank and other sources should continue. Identification by the government in this manner should give these firms an advantage in obtaining access to commercial loans as well. Overall, this Japanese industry is likely to continue its pattern of easier access to capital than the American industry.

- Barriers to Japanese entry into the systems integration of airframes, engines, and avionics remain. The cost of maintaining and extending systems integration capabilities through new programs has increased for U.S. companies, but the price of entry—through acquisition or accumulating expertise—is likely to fall, driven by excess industry capacity. An industry increasingly characterized by global partnerships and programs will allow the Japanese to continue building a revenue and technology base to make the jump to systems integration at an opportune time after the turn of the century. *Perhaps the most significant barriers to this jump are related not to inherent capabilities, but to the perception of risks in the market for a "Rising Sun" jet and potential impacts on U.S.-Japan relations.*

- It is also important to remember that the Japanese aircraft industry currently faces its own problems and challenges. The rapid appreciation of the yen during 1993 and the aircraft slump have dealt a double blow to the aircraft divisions of the heavies. Japanese government and industry are currently contemplating how to redeploy resources to build the industry in the future. In addition, countries such as South Korea and Taiwan have formulated national strategies to build aircraft industries through international alliances in the same manner that Japan has. In order to remain a force in the global industry, it will probably not suffice for Japan to stand in place, and vision will be required to move ahead.

POSSIBLE SCENARIOS AND IMPLICATIONS FOR
U.S.-JAPAN TECHNOLOGY LINKAGES

All of the possible scenarios outlined here assume that the Japanese industry will be a source both of valuable technological ties for American firms and potential or actual competitors.

Rough Continuation of Current Trends

Even under the most optimistic circumstances, the global aircraft industry will be depressed for several more years. The U.S. industry will shrink further and consolidate. In coming years it will be increasingly necessary for U.S. aircraft manufacturers to continue to develop and invest in high-quality, low-cost manufacturing capability, as well as access Japanese manufacturing technology and stay ahead in product and design technology. Maintaining a broad-based supplier network is also critical. The danger during the current restructuring is that U.S. companies in some areas of Japanese strength will completely exit the industry. Japanese direct investment in some of these areas (materials) is already occurring and would not raise the national security concerns comparable to the acquisition of a prime contractor. Japanese companies are not awash in cash right now, but they could probably come up with sufficient funds for strategic acquisitions that make business sense (especially since the yen rate makes it attractive to move some manufacturing out of Japan anyway). U.S. acquisitions could also occur in areas where Japanese companies are beginning to establish a higher profile (engine components).

In the case of complete U.S. exit from certain key areas of the aircraft supply chain in which Japanese companies are strong, there would perhaps be no U.S. companies with an incentive to obtain Japanese technologies in those fields (except the primes in some cases). *In areas where restructuring leaves one or more U.S. companies in a position to compete with Japanese firms, the ability to make adequate long-term investments in equipment and R&D will become a critical imperative.*

One particular concern is whether U.S. aircraft primes and suppliers will be able to maintain control over their crown technological jewels in this harsh environment. It is possible that in order to survive, some companies will be tempted to make large-scale technology transfers that enable foreign industries (including, possibly, the Japanese industry) to compete more effectively with the United States. The future of the Committee on Foreign Investment in the United States (CFIUS) process (which reviews foreign acquisitions and provides a mechanism for blocking them when they endanger national security) is unclear, and the recent trend has been toward a relaxation of export controls. Therefore, barriers to the outflow of significant U.S. aircraft technologies may be lowered in coming years.

Asian Airbus

Some analysts discuss the possibility of an Asian Airbus, particularly if one of the existing airframe primes exits the business. Some believe that this possibility has been enhanced in recent years by Japan's industrial cooperation and aid policies in Asia.

Although possible, a Japanese-led Asian aircraft consortium would be very difficult to put together. The Japanese have shown reluctance to make the large, risky investment needed to enter the market as a prime integrator, and this strategy would appear to be the most expensive and risky of them all. Japan would be the logical country to lead a viable Asian aircraft consortium, but it is hard to conceive of China, Korea, and Taiwan (countries with large aircraft markets) rushing to sign up. This would leave a Japan-led consortium vulnerable to a number of counterstrategies. In fact, Japanese press reports have speculated about an Asian aircraft consortium that would exclude Japan.[8]

Asia/Japan Cooperation with Airbus

Still another possibility would be an alliance involving Airbus and Asian aircraft industries. Airbus has been courting Japan in recent months, and newly industrializing countries in Asia are anxious to promote domestic industries. Although it seems unlikely that all of these countries would team with Airbus (and jeopardize linkages with U.S. companies and traditional relationships with the United States), a group of them might be stimulated to do so. The issue of reallocating Airbus work share to make room for the Asian partners could also prove to be a stumbling block to such an alliance.

Japan Squeezed

Especially if Japan makes aggressive moves to increase its global presence (by making a major acquisition at the integrator level, launching an independent program, aggressively playing both sides against the middle in its international alliances, or other circumstances), it is possible that the Japanese heavies in particular could be squeezed by companies from other nations following a similar strategy. The Japanese have moved quite far down the experience curve in making aircraft structures, but the main requirements necessary to do that sort of work are general manufacturing excellence and lots of patient capital. It would not be impossible for Korea or Taiwan (or others who have made it known they are available) to emulate Japan, particularly if one of the major airframe manufacturers has a strong incentive to put one of these countries in business. However, this scenario is less likely to affect some materials and

---

[8]See "YSX Keikaku in fuan no tane" (Seeds of Doubt for YSX Plan), *Nihon Keizai Shimbun*, October 14, 1993, p. 11.

components manufacturers (such as Toray and Hosiden) because of their existing strong global market positions.

Russian Wildcard

The Russian aircraft design bureaus possess considerable design capabilities. Although it appears that the Russians have been quicker to team with U.S. companies than with firms from Japan, over the long term the combination of Russian integration and design skills with Japanese capital and manufacturing know-how would seem to represent a potentially powerful combination. While the two industries have agreed to launch some small-scale collaborative activities, considerable obstacles remain to a smooth Russian-Japanese working alliance in aircraft design and manufacturing.

Resurgent U.S. Industry

Even in the current tough business climate, some U.S. companies are making the long-term investments in technology and R&D necessary to retain leadership in this industry. The concern is that if current trends continue, many U.S. companies will not make these investments and large segments of the U.S. industry will face severe challenges to their survival. However, the opportunity now exists to take steps—at the corporate and national levels—that can change these trends and enable the United States to retain its technological strength, maintain a full-scale manufacturing base, and compete strongly in world markets on the basis of superior technology, design, and manufacturing performance.

By introducing advanced technologies into new aircraft while lowering manufacturing costs, U.S. companies can take advantage of continuing upheavals in the global industry to reenergize their leadership. The keys to creating conditions in which a resurgence of U.S. leadership in aircraft manufacturing can take place are outlined in the following chapter.

5

Conclusions And Policy Recommendations

THE GLOBAL CONTEXT AND U.S. NATIONAL INTERESTS

Aircraft manufacturing is critical to America's long-term economic growth and national security. In terms of the economy, it is a major factor in domestic employment and international trade; in terms of security, U.S. airpower has played a major role in strategic nuclear deterrence, and the Gulf War clearly demonstrated the importance of modern, technically advanced aircraft to America's military superiority. Additionally, this industry is global—not only in its markets and its basic mission, but also in its industrial structure, technical talent, and financing. Finally, aircraft development requires enormous capital investments (tens of billions of dollars) whereas payback is achieved only over the long term and individual programs face a high risk of never breaking even. It is this combination of factors that makes the aircraft industry both unique and of significant national importance. Thus, it has historically been singled out for government support—particularly through advanced research funding by the National Aeronautics and Space Administration (NASA) and through R&D and production funding by the U.S. Department of Defense (DOD).

Today, the U.S. aircraft industry remains a world leader, but significant adjustments will be needed for it to remain so in the future. Over the next decade, U.S. industry will continue to come under increasing international competitive pressure. The aircraft industries of Europe, Japan, Russia, Taiwan, China, and other nations are aggressively seeking opportunities to tap into the expected long-term growth in commercial aircraft markets. Forecasts for growth in the Asian market are particularly impressive. Heightened international competition will take place in an environment of unprecedented U.S. industry restructuring as a result of dramatic reductions in the defense budget. Therefore, U.S. industry will be severely challenged over the next decade just to hold its current position in global aircraft manufacturing. Achieving growth in global market share will be an even more difficult task.

This study of U.S.-Japan alliances illustrates the key features of this evolving global competitive environment and highlights the broad challenges faced by the U.S. aircraft industry. In order to reenergize U.S. leadership in the face of these challenges, a new approach must be developed by industry and government.

Conclusion

- *Leadership in aircraft design and manufacturing—including a full spectrum supply chain—remains a vital U.S. national interest.* In order for the United States to maintain its leadership position in this critically important industry, it is essential that aircraft be singled out for specific, strong, government-industry partnering in the development and implementation of a long-term strategy.

THE JAPANESE AIRCRAFT INDUSTRY

Japan is currently a significant player in global aircraft manufacturing. Japanese companies are formidable competitors in a number of aircraft subsystem and component areas. Although Japanese industry is not competing today at the prime integrator level, Japan already possesses or could acquire the capabilities needed to do so. The committee has seen that Japan is making the long-term investments necessary to be a world leader in air transport design, development, and manufacturing. Japan's primary strength lies in the manufacturing capabilities of its companies, and Japanese firms are focusing on low cost and high quality as differentiating factors.

Japan has established an aircraft industry as a matter of national policy with managed internal competition but with a resilience to changing economic conditions. Technological, financial, and human resources are leveraged across civil-military, supplier-prime, and horizontal interfaces to maximize industry's long-term competitive position. Strong industry-government partnership in formulating and implementing strategies in the aircraft industry has long been a key feature of the Japanese environment. These characteristics of the Japanese aircraft industry will serve it well as it seeks to expand its global presence in the post-Cold War competitive environment.

Japan is committed to deepening its capabilities across a range of aircraft-related technologies and to increasing its presence in the commercial aircraft market.[1] Japan's current strategy is to develop international linkages to achieve these goals. Japan has more options today in terms of international linkages than it has ever had (Russia, Europe, etc.), and U.S. government and industry should not assume that they have a lock on the action.

Conclusion

• *Japanese industry's role in global aircraft manufacturing, design, and technology development will continue to grow.* Japanese industry retains an option to enter the market as a prime integrator and/or to further expand the scope of its international alliances. Although currently facing difficulties, the Japanese industry has inherent strengths that will see it through the current downturn and allow it to emerge as a more formidable competitor in both established and emerging areas.

U.S.-JAPAN TECHNOLOGY LINKAGES

The 40-year modern history of cooperation between the United States and Japan in the aircraft, and associated subsystem, industries has been mainly positive for both sides. Japan has used linkages to build its technological and manufacturing capabilities in military and commercial aircraft production. American industry has earned significant revenues from Japan through aircraft sales and licensing, as well as the benefits of effective business partnerships.

Characteristic linkage mechanisms have evolved from licensed manufacturing of U.S. designs to the present stage in which a number of alliances involve the design of significant components and subsystems by Japanese aerospace companies. We now appear to be entering a new stage, partly spurred by several Japanese initiatives, of more extensive cooperation in major aerospace R&D programs (such as the FS-X and HYPR).

U.S.-Japan linkages can continue on a constructive basis, provided there is balance and fairness in the flow of technologies. Here, the challenge for the United States is to continue to build effective U.S.-Japan relationships, while

---

[1]Recent news reports on renewed efforts to line up partners for a YS-X feasibility study and Mitsubishi Heavy Industries' (MHI) participation in Bombardier's Global Express business jet program (MHI will manufacture the wings and center fuselage) underscore the intention of Japanese industry and government to push forward in aircraft despite tough economic times. See Christopher J. Chipello, "Bombardier Board Approves Plans for Corporate Jet," *Wall Street Journal*, December 20, 1993, p. A5; and "YSX, Nichi-Bei-O-Chu de Kaihatsu" (YSX to Be Developed by Japanese-U.S.-European-Chinese Partnership), *Nihon Keizai Shimbun*, December 28, 1993, p. 1.

developing a sharper focus on defining the technologies that we want to flow to U.S. industry, as well as those that we should maintain domestically.

In maintaining its leadership and meeting the global competition, the overall U.S. approach must not be "protectionist" or "defensive" but proactive—to maintain U.S. leadership and enhance U.S. capabilities. Markets and technology development in this industry are global—Japan and the rest of Asia are of increasing importance in both areas. U.S.-Japan and other international technology linkages are facts of life and likely to increase globally in this industry. Efforts must be made to structure alliances with Japan so that they enhance U.S. access to Japanese technology, markets, and capital.

Conclusion

- *The challenge for U.S. industry and government is to stay ahead,* using technology linkages with Japan as part of a strategy to build capabilities needed for a strong domestic manufacturing and technology base and an industry consistently capable of effective global competition.

DEVELOPING A U.S. STRATEGY

The majority of the actions needed to maintain America's leadership position in the aircraft industry and to ensure mutually beneficial relationships between American and Japanese firms must be the responsibility of the U.S. aircraft industry itself—both prime integrators and the supplier base. However, it is also clearly necessary for the U.S. government to create a favorable overall environment for these actions, as well as play a specific role in creating incentives or actually making selected, limited investments. Providing the desired overall environment and assistance requires both a long-range strategy and an institutional structure to implement it. Currently, neither the strategy nor the needed institutional mechanisms exist.

The committee therefore recommends a five-part strategy to address U.S.-Japan relationships and the larger competitive challenges facing the U.S. aircraft industry. The objectives are to revitalize U.S. aviation leadership (both in technology and in market share) and to maintain a significant, full-spectrum domestic engineering and manufacturing base. The five elements of a comprehensive U.S. strategy include:

1. maintaining U.S. technological leadership,
2. revitalizing U.S. manufacturing capabilities,
3. encouraging mutually beneficial interaction with Japan,
4. ensuring a level playing field for international competition, and
5. developing a shared U.S. vision.

Maintaining U.S. Technological Leadership

U.S. leadership in aviation is largely the result of a continuous, long-term stream of investment that has supported the development of a wide range of advanced technologies. This investment has come from the private and the public sectors. The current massive restructuring on both the military and the commercial sides of the aircraft business makes it critical that U.S. technological leadership be maintained. Industry, NASA, and DOD all have a vital role to play; and other agencies (Department of Transportation/Federal Aviation Administration, Department of Energy, Office of Science and Technology Policy, National Economic Council) also are significantly involved.

Clearly, aviation is an area in which the best policy for future U.S. economic growth is to stay ahead. Other nations, including Japan, Europe, Russia, and China, are focused on this industrial sector, and U.S. industry's technological lead has narrowed considerably in recent years. For some time the civilian aviation arena has been characterized by incremental technological advance, rather than by dramatic breakthroughs comparable to the high-bypass engine. Partly for this reason, aircraft manufacturers currently face intense price pressure. The Japanese have recognized this trend and are now focusing on low-cost, high-quality manufacturing as a differentiating feature. For U.S. industry to survive and grow in this environment, the United States should take steps to overhaul and refocus aircraft-related R&D activities.

What the United States must do is strive for quantum improvements in the application of process as well as product technologies. This will involve setting and meeting concrete targets—such as lowering the unit and life-cycle costs of aircraft and of air travel for advanced subsonic aircraft and the next-generation high-speed transports by one-third or more. The entire system will need to be addressed—from the cost of aircraft and engines, to fuel efficiency, maintenance, reliability, and airport air and ground operations.

This will require a significant restructuring of the large R&D investments government makes—mainly through NASA and DOD—in order to achieve greater efficiency and commercial impact. The committee supports NASA's recent initiatives to increase research on subsonic aircraft and propulsion systems. NASA should continue on this course by aggressively increasing its emphasis on developing cost-effective, product-applicable technologies, increasing the flow-through of R&D funding to industry, and supporting greater cooperative efforts among U.S. companies. Particularly in the subsonic area, an important focus should be on lowering the cost to industry of incorporating new technology in aircraft and related systems. Although greater attention and resources should be devoted to advanced subsonic aircraft, NASA's partnership with industry in high-speed civil transport research should also continue as a high priority.

The Department of Defense aircraft R&D budget for enabling technologies must be maintained at current levels despite overall cuts in the defense budget (see Figures 2-2 and 2-3). DOD should also reorder its procurement and R&D funding priorities to promote integration of military and civilian systems. As in the case of NASA funding, more emphasis should be placed on cost-effective technologies. The issue of civil-military integration is treated in more detail below in the discussion of revitalizing U.S. manufacturing capabilities, but this goal should be a focus of Department of Defense R&D spending as well. The committee is encouraged by recent initiatives[2] and the stated positions of DOD officials, but the barriers to changing old ways of thinking and doing business should not be underestimated. Some aspects of the Japanese experience are instructive. For example, working with fly-by-wire technology on the T-2 trainer helped maintain Japanese industry's strong position in primary actuation, and Japanese strength in microwave integrated circuits for civilian applications contributed to the development of the FS-X phased array radar.

In addition, U.S. industry must continue to invest its own resources in new technology development. For many aircraft and aerospace companies, this is a difficult prospect in the current environment. Even companies that now have a healthy cash flow may be reluctant to make long-term investments because of uncertainties related to ongoing industry restructuring. In addition, the recently renewed R&D tax credit does not serve as an incentive for companies shifting from military to commercial applications unless the overall amount of R&D spending increases. By working closely with industry in its technology development programs and modifying the tax credit, government could help industry maintain the necessary level of R&D investment.

Finally, cutting-edge academic research in fields such as computational fluid dynamics also makes a substantial contribution to U.S. capabilities. A number of government agencies fund relevant academic research (NASA; DOD through the Advanced Research Projects Agency, Air Force Office of Scientific Research, and Office of Naval Research, and the National Science Foundation). Currently, there is no coordination of this investment across agencies. In recent years the federal government had begun to coordinate some of its technology activities through the Federal Coordinating Council on Science, Engineering and Technology, an initiative that the current Administration has consolidated with other interagency policy councils, establishing the National Science and Technology Council. The High Performance Computing and Communication Initiative is another good example. Agencies funding aeronautical research at universities should establish a similar committee that incorporates industry input in order to achieve a better focus on work relevant to industry.

---

[2]An Advanced Research Projects Agency initiative on "low-cost aircraft" and an Air Force initiative on "lean aircraft manufacturing" are recent examples.

Recommendations

- The 35 percent increase in NASA aeronautics R&D funding for fiscal 1994 is a step in the right direction, and efforts should be made to continue this percentage increase for the next three years, primarily through reallocation within NASA.[3] NASA should further expand its enabling technology R&D programs in subsonic aeronautics and propulsion systems, with the primary objective of reducing the initial investment and operating cost of future aircraft and subsystems.

- NASA's traditional role in basic research should be expanded to include nearer-term, product-applicable technologies. This will involve support for more technology demonstrations aimed at lowering the cost to produce, operate, and support aircraft incorporating new technology.

- NASA should significantly increase the share of aeronautics funding contracted to industry (currently 17 percent) with the objective of reaching 50 percent over the next five years, in particular, targeting technologies relevant to suppliers.

- DOD should maintain the current level of R&D support allocated for the development of advanced "enabling" technologies for the aircraft industry at both the prime and, particularly, the subcontractor levels.

- The committee supports other groups that have called for the R&D tax credit, which was recently extended for two years, to be made permanent.[4] It also believes that a mechanism should be developed to avoid penalizing companies that reorient their R&D from defense-unique to dual-use or commercial areas. The focus should be on creating incentives for greater commercial and dual-use R&D investments, even if the level of defense R&D is reduced.

- An interagency body should be created to coordinate—with industry cooperation—federal government investment in university and national laboratory research in aerodynamics and other related fields (e.g., computer science and materials science).

Revitalizing U.S. Manufacturing Capabilities

In view of the global competitive environment of continuing cost pressures on aircraft manufacturers, U.S. primes and suppliers will have to continually improve manufacturing performance in terms of cost, quality, and delivery to remain competitive. This is especially critical in light of the large investments in state-of-the-art equipment being made by the Japanese aircraft industry. While some U.S. companies are making the necessary investments, many are

---

[3]By using the FY 1994 authorization of $1.69 billion as a baseline, this would imply an aeronautics budget of about $4 billion in FY 1997.

[4]See National Science Board, *The Competitive Strength of U.S. Industrial Science and Technology: Strategic Issues* (Washington, D.C.: U.S. Government Printing Office, 1992), p. 46; and Competitiveness Policy Council, *A Competitiveness Strategy for America* (Washington, D.C.: U.S. Government Printing Office, 1993), p. 19.

finding it difficult because of the current commercial aircraft slump and defense cutbacks. Although an aircraft industry structure with fewer U.S. players at various levels is perhaps inevitable, it is important that the remaining companies—particularly the supplier networks—have the wherewithal to match or exceed the manufacturing performance of Japanese and other international competitors.

There are four major aspects to ensuring that the U.S. aircraft industry develops world-class manufacturing capability. First, companies themselves must make the necessary investments in new equipment. Other groups have called for the reintroduction of an investment tax credit.[5] Although the committee is well aware of the severe budget environment, its view is that a tax incentive structured to encourage companies in this capital-intensive industry to stay on the cutting edge of manufacturing technology would be in the national interest.

The second requirement for revitalizing U.S. manufacturing capabilities in aircraft is greater civil-military integration to increase the economic impact of DOD aircraft spending. This should be a priority in DOD aircraft R&D, production, and support investments. DOD spending has a pervasive influence on the business activities of aircraft companies—including investment and manufacturing. Over time, there has been a dramatic widening of the gap between military and civilian aircraft R&D, engineering, and production within firms in the United States. This is apparent in the major technical issues that have become the focus of military aircraft development over the past several decades, such as stealth and high maneuverability, which have little commercial relevance. Although DOD will continue to have some unique requirements, efforts to increase the dual-use applicability of defense systems and components wherever possible would lower procurement costs and support the commercial competitiveness of defense contractors.

The benefits to industry of a more dual-use oriented defense industrial base might be greatest in terms of manufacturing. For example, both Boeing and McDonnell Douglas have found it prudent to separate military and commercial transport work because of the significantly higher administrative and other costs associated with defense contract work. The potential benefits of removing these barriers are likely to be even greater in the supplier base. Thus, DOD must provide incentives for companies to integrate their military and commercial production, and reduce the huge barriers to civil-military integration—including cost-accounting standards, military specifications, procurement practices, and rebalancing DOD's stress on performance versus cost. The Japanese aircraft industry is achieving plant integration in its advanced aircraft manufacturing operations—in areas such as composites and metal parts—for the FS-X fighter and the 777 transport. The committee commends recent

---

[5]Council on Competitiveness, *Technology Policy Implementation Assessment 1993*, (Washington, D.C.: Council on Competitiveness, 1993), p. 6.

statements by DOD officials indicating that tackling this problem is a top priority, and urges timely and vigorous follow-through.

A third requirement for ensuring a strong U.S. aircraft manufacturing base is building more effective vertical relationships between firms at all levels of the supply chain. The importance of these relationships for advancing state-of-the-art manufacturing is obvious in some areas—such as the relationship between structures manufacturers and machine tool builders. Effective supplier relations can significantly improve design and manufacturing performance in terms of cost, quality, and cycle time throughout the aircraft manufacturing infrastructure. This is one of the key strengths of the Japanese aircraft industry.

Most of the responsibility for building vertical links that improve the performance of the U.S. aircraft manufacturing system lies with the companies themselves. There is evidence that a number of U.S. primes and suppliers are making positive changes—particularly on the commercial side of the aircraft business, where some primes are increasingly recognizing the long-term benefits of closer, more stable links with suppliers and are instituting programs that help increase supplier capabilities. However, policy changes should create incentives that support and expedite this process, particularly on the defense side. Although the Department of Defense has made efforts to encourage more effective relationships, further changes in R&D and procurement funding contract mechanisms could encourage closer cooperation between U.S. firms and their suppliers, leading to lower costs and greater technology sharing in the long run. The occasional practice of "breaking out," or putting the supply of parts for ongoing programs up for international competitive bids, has an especially adverse impact on the supplier base. R&D and procurement funding and contract mechanisms should encourage, rather than discourage, cooperation among U.S. companies.

The fourth aspect of revitalizing U.S. aircraft manufacturing is R&D. Of the additional funding for aeronautics and aircraft-related R&D recommended above by the committee, a significant portion should be devoted to process technology development. DOD's Mantech program is an existing mechanism that a number of U.S. companies have found delivers significant benefits.

Recommendations

• *In order to maintain global leadership, U.S. aircraft manufacturers—both primes and suppliers—must invest in high-quality advanced manufacturing processes* that will position them to compete as low-cost, high-quality, low cycle time producers in the years ahead. Introduction of a tax incentive for productivity-enhancing investments should be studied. The tax credit should be targeted to investments by lower-tier suppliers in technologies considered critical, or to investments in advanced manufacturing equipment and training.

• The Department of Defense should reform the procurement system to promote greater civil-military integration. This reform should include more extensive use of commercial item descriptions, greater emphasis on low cost and high quality in addition to performance, provision of data bases and training to enhance the use of commercial specifications,[6] and increased use of commercially available components and processes. Perhaps most important, DOD should reduce, to the extent possible, barriers to utilizing common manufacturing facilities for military and civilian aircraft production through revision of its accounting and oversight requirements, military specifications and standards, and procurement practices.[7]

• The committee concurs with recent Defense Science Board recommendations on low volume production and further recommends that DOD explore steps toward revitalizing U.S. aircraft manufacturing capability, such as carrying prototype aircraft systems and subsystems forward in limited quantity fabrication in order to demonstrate low-cost "manufacturability" in addition to specified performance.[8]

• The Department of Defense should modify its procurement and R&D funding mechanisms to eliminate current disincentives for long-term prime-supplier relationships that enhance quality and lower costs.

• As part of a stronger emphasis on technologies that are applicable in the near term and contribute to lowering aircraft and air travel costs, NASA and DOD aircraft-related R&D programs should place a high priority on manufacturing and design processes.

Encourage Mutually Beneficial Interaction with Japan

The committee recognizes that U.S.-Japan alliances that transfer or develop technology are a fact of life in this industry. Cooperation with Japan has delivered significant benefits to the U.S. aircraft industry over the years, which have been outlined above in the discussion of distinctive features of U.S.-Japan linkages. Yet technology transfer to Japan also carries risks for individual companies and U.S. industry as a whole. In addition, the environment surrounding U.S.-Japan linkages has evolved significantly. Japan's growing technological and manufacturing capabilities, as well as the passing of the Cold War context in which alliances were structured in the past, necessitate a new approach by the United States to ensure that the benefits of cooperation are maximized and the risks are managed.

U.S. government and industry must create new mechanisms and devote additional resources to encouraging mutually beneficial U.S.-Japan relationships in several key areas: (1) technical information management and technol-

---

[6]See Center for Strategic and International Studies, *Integrating Commercial and Military Technologies for National Strength* (Washington, D.C.: CSIS, 1991) for a detailed analysis of more specific options on specifications and fostering military-civilian synergies.

[7]For example, DOD's Manufacturing Quality Requirements have been revised recently to permit integration of civilian and military production in the electronics area.

[8]See Defense Science Board, *1991 Summer Study on Weapon Development and Production Technology* (Washington, D.C.: Office of the Under Secretary of Defense for Acquisition, 1991).

84

ogy benchmarking; (2) identification, valuation, and control of critical technologies; and (3) education and training. Although Japan-related activities may require special attention and greater resources, it should not be the sole focus of this new approach—the U.S. aircraft industry's technology linkages are global, and greater efforts to maximize the benefits of international cooperation should reflect this.

Information Management and Technology Benchmarking

Competitiveness in high-technology industries such as aircraft depends to a great extent on how quickly products can be brought to market. Reducing design and production time requires an efficient use of resources, both human and technical. The U.S. technology infrastructure is loosely linked, and initiative has a strong "bottoms-up" orientation. This structure promotes innovation but often inhibits the diffusion of technology.

The U.S. government, by virtue of its broad information collection capabilities, is in a unique position to gather, package, and disseminate useful technical and business information from global sources to U.S. industry. Although this is an appropriate general role for government, collection, coordination, and dissemination efforts require a stronger industry focus than they have received up to now. Aeronautics could be a test case for a new approach. The U.S. government already has several programs that are relevant to this effort. For example, the Japan Technology Evaluation Center (JTEC) overseen by the National Science Foundation (NSF) has conducted numerous studies in recent years that benchmark Japanese technologies. Several of these studies have been funded by NASA or DOD and have examined aerospace-related technologies such as advanced composites and supersonic/hypersonic propulsion.

In addition to existing programs with relevance to Japan, other U.S. government agencies collect, translate, and disseminate a variety of technical and business-related information of potential use to the U.S. aircraft industry.[9] What is required is a coordination mechanism with strong industry input that focuses these efforts. The most logical place for this coordination function is the Technology Administration of the Department of Commerce. This effort should ensure that U.S. government information activities are tied to U.S. industry needs, and should permit a broader spectrum of U.S. industry to access usable knowledge on Japanese aircraft technologies and industry activities.

The committee believes that it is also necessary, as part of this new effort, to support U.S. industry access to information through an industry outpost in Japan. Although most of the large U.S. aerospace companies maintain a presence in Japan, and the American Aerospace Industry in Japan (AAIJ) repre-

---

[9]Another relevant activity is the information exchange being launched between NASA and Japan's National Space Development Agency. See Laurie Harrison, Glenn P. Hoetker, and Thomas F. Lahr, "Access to Japanese Aerospace-Related Scientific and Technical Information: The NASA Aerospace Database," *Japanese Technical Literature Bulletin*, June 1993, p. 1.

sents their common interests, many smaller U.S. companies do not have the resources to maintain a Tokyo office. Even the large companies focus their Japan efforts primarily on marketing rather than on technology access and related issues. A U.S. government-funded industry outpost in Japan—directed by a technically trained American fluent in Japanese—could serve as a source of "on the ground" information and as a liaison to Japanese government and industry. For example, some Japanese government advisory committees (*shingikai*) have invited foreign participation in recent years. The U.S. industry liaison would be available to participate in—or at least track—these advisory activities.

Another area in which new approaches to dissemination are needed is information on the flowback of Japanese technical improvements and indigenous Japanese technologies through specific U.S.-Japan military aircraft programs—most notably the FS-X. The Department of Defense and the Department of Commerce are already making a contribution in this area, but additional efforts are necessary. For example, the wide range of opinions expressed about the value of FS-X technologies such as the composite wing and advanced avionics to U.S. industry illustrates the need for a more systematic effort. Sufficient resources should be made available for the responsible agencies to catalogue and distribute flowback data on an industry-wide basis.

Identification, Valuation, and Protection of Critical Technologies

U.S. government involvement in technology linkages between U.S. aircraft manufacturers and international partners has occurred mainly in the context of the export control system. U.S. government involvement has been extensive where government-to-government agreements on bilateral or multilateral military programs are required (F-15 licensed production and FS-X codevelopment with Japan; F-16 coproduction with several European nations). Even for purely commercial links, export licenses are sometimes required because of the dual-use character of the technologies involved (Boeing's joint work with Japan on the 7J7), and in rare cases, alliances have been the subject of discussions at the highest levels of government (the formation of CFM International).

The future U.S. government role in international transfers of aircraft technology is unclear. There has been a relaxation of export controls resulting from the end of the Cold War. The committee believes that while this is mainly a positive trend, it is important that the United States, for national security reasons, retain export controls on a limited number of critical aerospace technologies.[10]

The committee further believes that outside the few identifiable critical areas covered by export controls, actual negotiations for technology transfer and

---

[10]A longer-term reorganization of the system to increase clarity and user friendliness may be necessary, as outlined in the National Research Council report *Finding Common Ground—U.S. Export Controls in a Changed Global Environment* (Washington, D.C.: National Academy Press, 1991).

other international collaboration are the responsibility of U.S. companies (or teams of U.S. companies) negotiating with prospective Japanese partners or the Japanese government. Companies themselves are normally the best judges of what relationships will serve their own long-term growth and interests, but national and corporate interests sometimes diverge. In addition, the long-term impacts of technology transfers are often difficult to predict. In the U.S.-Japan context, the semiconductor industry provides a useful example. U.S. companies licensed the key basic technologies for semiconductor manufacturing to Japanese industry in the 1960s, under conditions in which access to the Japanese market was severely limited.[11] It was not until years later that the Japanese semiconductor industry developed into a formidable competitor.

The U.S. aircraft industry has often transferred technology abroad in order to enhance market access, in many cases without a commensurate return technology flow. This historical pattern is understandable given the large gap in technological capabilities that often existed between U.S. industry and most foreign partners. However, in the current environment of intense global competition, U.S. industry cannot afford to ignore the increasing sophistication of its overseas partners and must aggressively pursue a balanced flow of technology wherever possible. Especially in the current business context in which international alliances are a fact of life, U.S. industry must make the best deals possible. Yet what distinguishes "good deals" from "bad deals"? Although even experts might differ over particular cases, the accumulated knowledge and expertise of U.S. industry regarding technology transfer in international alliances constitute a valuable resource. Past experience and current imperatives suggest the need for an independent body to develop guidelines for technology transfer consistent with national interests.

More systematic consideration should be given to identifying and protecting critical aeronautical technologies at both the company and the industry levels. The committee believes that a most promising approach is to establish a new mechanism aimed at drawing on and disseminating the accumulated knowledge of industry and other experts. This effort led by the private sector could be a valuable resource for companies as they negotiate international technology alliances, the ultimate goal being to expand the data base required to properly value corporate technological assets and to structure international cooperation that brings clear economic and technological benefits to the United States. This could be accomplished by a working subgroup of a new National Aviation Advisory Committee, which is described in greater detail below. This activity ideally should incorporate the following tasks:

1. Publish and periodically update a description of the critical technologies for the aircraft industry, to be used as an informal input for company decision-

---

[11] See National Research Council, *U.S.-Japan Strategic Alliances in the Semiconductor Industry* (Washington, D.C.: National Academy Press, 1992), pp. 11-12.

making and for government R&D funding and international benchmarking activities.

2. Develop guidelines covering the international transfer of commercial aerospace technology—including the development of data and methods for valuing technology—that would help corporate managements make and objectively evaluate technology transfer decisions.

3. Carry out periodic assessments of international technology transfers and measure progress toward increasing the flow of aerospace technology into the United States, including acquiring the data needed to undertake these assessments.

Education and Training

American universities and research institutions play mainly a background role in U.S.-Japan linkages in the commercial aircraft industry. Nevertheless, this role can have a crucial impact on moving toward more productive and balanced technology linkages. This occurs primarily through the training of American engineers, scientists, and managers in Japanese language and area studies. Educational programs at American universities constitute an important vehicle for building awareness of Japan as a global competitor, and for providing students with the information and skills needed to interact with the Japanese in a more productive way (presumably enabling an enhanced flow of technology from Japan to the United States).

A number of university programs have been established in recent years to train young scientists, engineers, and other professionals in Japanese language and technology management. Three years ago, DOD (through the Air Force Office of Scientific Research) launched a mechanism to fund such programs, in order to increase the effectiveness of those already existing and to spur the formation of new centers.[12] As a result, a larger pipeline exists for training technologists and managers to operate effectively in a Japanese environment. Several U.S. aircraft companies are hiring graduates of these programs. Closer interaction between industry and university programs of this type would lead to mutually beneficial impacts, including employment of qualified graduates, internships for students, and training programs tailored to industry needs.

Education and training also have a bearing on technology outflow. In addition to technology transferred to overseas partners through licensing agreements, some technology flows out inadvertently as a result of inexperience or lack of training. Many employees do not realize that valuable technology can be transferred in a casual conversation or in activities such as a presentation to a professional society. Nor do they realize that their company's commercial technology may be specifically targeted by their foreign competitor or would-be

---

[12]National Research Council Committee to Assess U.S.-Japan Industry and Technology Management Training Programs, *Interim Report* (Washington, D.C.: National Academy Press, 1993).

competitor. "Ego" also comes into play as engineers try to impress their foreign contacts with their knowledge and accomplishments, a characteristic encouraged by potential foreign competitors eager to acquire technology.

It is in the general interest of U.S. industry to lessen this inadvertent technology outflow. Training programs and processes patterned after those used to protect DOD classified technology and information should be instituted at an industry level for companies establishing international alliances. The objective of the training and processes would be to make all employees knowledgeable of what type of technology should be protected and what they must do to protect it. In addition, employees who are going to interface regularly with foreign competitors should, whenever possible, learn the language of their counterparts. This could be augmented by Japan-specific training for negotiations and technical interaction—perhaps instituted as an industry outreach activity by one or more of the university-based Japan technology management centers.

Recommendations

• The Department of Commerce should consider using the aircraft industry as a test case for a new approach to coordinating information collection and dissemination activities in various agencies, the goal being to increase the utility of government information to industry. This effort should incorporate regular technology benchmarking and include the establishment of a small aircraft industry outpost in Japan.

• The Departments of Defense and Commerce should devote additional resources to a systematic program of cataloguing, evaluating, and disseminating to industry information about technology flowback and indigenous Japanese technologies in connection with the FS-X and other collaborative military aircraft programs. An important goal of this effort should be to establish a basis for making judgments about the potential value of this technology to DOD and U.S. industry, and to improve U.S. access to Japanese manufacturing technology when there are both a demonstrated U.S. need and potential users.

• One of the central tasks of a new National Aviation Advisory Committee should be to support U.S. industry decision making in the areas of critical technologies and international technology transfer. A working subgroup of the new committee should identify critical technologies, develop guidelines for the transfer of aerospace technology, and conduct periodic assessments of international technology flows.

• University-based programs that provide Japanese language and management training to young technologists and other professionals should strengthen interactions with the U.S. aircraft industry to help meet industry's need for managers and technologists who can interact effectively with Japanese counterparts.

• The U.S. aircraft industry should collaborate to develop a training program for employees involved in technology exchange to enhance protection of critical technologies and effective technology transfer (both inflow and outflow) where appropriate.

Ensuring a Level Playing Field for International Competition

Japan has no formal barriers to aircraft imports, and to this point its industry subsidies have not caused massive distortions of international markets. However, in light of heightened international competition in all segments of the aircraft industry and the inclination of governments to be heavily involved in the development of national industries, U.S. trade policy must be a key element in any U.S. strategy for the aircraft industry.

Continued U.S. leadership in aircraft requires that trade policy support fair global competition by limiting massive government subsidies to competitors. Although the issue was set aside in the agreement reached in the Uruguay Round negotiations of the General Agreement on Tariffs and Trade (GATT), it may be possible to gain agreement on multilateral trade rules that protect the interests of the U.S. aircraft industry. Goals for such an agreement include the multilateralization of last year's bilateral Agreement on Large Aircraft (which bans production supports and limits new program development financing) between the United States and the European Community, as well as a strong Subsidies Code agreement that applies to aircraft and provides for disciplines on export subsidies and a dispute settlement mechanism. Trade negotiations are particularly important in light of the emergence of new aircraft manufacturers not currently bound by all of the relevant GATT disciplines (Russia, China, and Taiwan). In formulating strategies for multilateral negotiations, the U.S. Trade Representative should work closely with industry.

Another aspect of supporting U.S. industry's position in international markets is the financing support of the Export-Import Bank. Export-Import Bank financing was very important to U.S. aircraft exporters during the 1970s, but its role declined during the 1980s. Over the past two or three years, Export-Import Bank guarantees have again become an important factor in the export of U.S. aircraft due to the deterioration in the financial strength of airlines worldwide. The U.S. government should ensure that Export-Import Bank guarantee and lending authority for aircraft exports is sufficient to meet sales opportunities.

Recommendations

• In order to support the position of the U.S. aircraft industry in international trade and ensure a level playing field, the U.S. government should strive through trade negotiations to achieve multilateral rules that will govern and reduce subsidies.

- The U.S. government should also maintain recently increased Export-Import Bank guarantee and loan authority to the extent needed to take advantage of export opportunities.

Developing a Shared U.S. Vision

The committee believes that the four elements of a U.S. aircraft industry strategy and the associated recommendations for action outlined above are necessary and stand on their own terms. However, the critical importance of this industry and the rapidly changing context demand ongoing high-level attention to these issues in order to ensure that a strategy is implemented. Continued American leadership in this industry also requires that the United States foster more effective working relationships within industry and between industry and government.

This study of U.S.-Japan aircraft linkages and the Japanese aircraft industry highlights the need for a new approach. The committee has seen how Japan's aircraft industry—both prime contractors and suppliers—works with government to maintain and constantly upgrade skills and technological capabilities. Despite the industry's small size and the fact that Japanese companies are not among the major global players in prime integration, Japanese aircraft manufacturers are well established as key suppliers in the global markets for commercial transports and engines, mainly as partners in programs led by U.S. primes. Over time, the level and sophistication of Japanese participation in these programs have steadily increased.

Strong, stable relationships between Japanese primes and suppliers ensure that technologies are diffused and benefit the entire aircraft manufacturing network. The Japanese aircraft industry currently faces a number of challenges as a result of civilian and military market contractions and exchange rate shifts, but the committee believes that the industry's demonstrated ability to function as a system will allow it to weather these shocks and emerge as a stronger global partner and competitor in the future.

International alliances, particularly those with U.S. companies, have played and continue to play an important role in the development of the Japanese industry. Collaboration on both the military and the commercial sides has been supported by the Japanese government and has been structured to ensure a steady flow of aircraft-related technologies from abroad, as well as to provide opportunities for Japanese companies to develop and enhance indigenous technological strengths through their program participation.

Although the U.S. aircraft industry has great strengths, and it would not be possible or desirable to duplicate the Japanese system here, contrasting our situation with Japan's highlights the challenges that we face and the critical need to cooperate and utilize resources more effectively. The United States spends a significant amount on aircraft-related R&D, yet its technological lead has narrowed in recent years. Although it would be impossible and counterproductive to ensure that every aircraft prime contractor and supplier remains viable during the current period of restructuring, we face a greater risk of losing some critical capabilities altogether because our prime-supplier relations are more arm's length and less extensive than Japan's.[13] Although larger U.S. aircraft companies have gained demonstrable business and even technological benefits from their relationships with Japan, the impacts on the supplier base have not been as beneficial, and the bargaining power of even the strongest U.S. companies has been affected by the necessity of competing with each other in negotiations with a coordinated Japanese industry.

As has been pointed out elsewhere in this report, the major challenges faced by the U.S. aircraft industry are broad and generic—current weakness in the global market for commercial aircraft, declining defense procurement, and tough competition from a range of established and new international players. Competition from Airbus is obviously immediate and significant in airframe integration, and other national industries such as Russia's may pose a challenge in the future. Although this assessment demonstrates the need for a shared vision for the U.S. aircraft industry, developing this vision will require a comprehensive approach.

In order for the U.S. aircraft industry to maintain a full spectrum of design, technical, and manufacturing activities in the United States and link itself more effectively with foreign economies, it will be necessary for U.S. stakeholders to find more effective ways of working with each other. In policy terms, this means that we need a mechanism to build consensus and implement strategy on an ongoing basis, as well as to remove unnecessary obstacles to cooperation that exist in the United States.

The committee considered several alternative mechanisms for developing a shared vision for U.S. aircraft industry development and for providing a continuing focus for the associated tasks identified above (developing investment and R&D incentives, identifying critical technologies, assessing international technology transfer, and developing guidelines for these transfers). One possibility would be to charge NASA or another existing agency with the task. Indeed, until it was reformulated as NASA and given responsibility for leading the space program, the main task of the National Advisory Committee for Aeronautics (NACA) was to perform R&D and provide research infrastructure to ensure U.S. leadership in aviation. In terms of the circumstances that exist today, the major disadvantage of reconstituting NACA, charging NASA with the task, or undertaking some other form of government reorganization are that more than a redirection in R&D policy is necessary, and the policy questions come under the purview of a number of agencies. Existing private sector committees such as the NASA Advisory Council or the Defense Science Board that advise individual agencies on their R&D programs could perform specific tasks

---

[13]For a comparison of the approaches taken by Japanese and U.S. aircraft suppliers in the face of periodic downturns, see David Friedman, *Getting Industry to Stick: Creating High Value Added Production Regions in the United States* (Cambridge, Mass.: MIT Japan Program, 1993).

related to developing a U.S. vision. However, these existing advisory committees share with their corresponding agencies a lack of breadth that constitutes a compelling reason for not designating one of them as the focal point.

Another alternative would be for the Aerospace Industries Association (AIA) to play a leading role, perhaps in combination with other industry associations. AIA conducts ongoing technology road map activities for the industry and can be expected to make further contributions toward addressing the issues raised in this study. There are, however, other factors that argue against an industry association serving as the focal point for formulating a shared vision and undertaking the associated tasks. In addition to incorporating a wide range of industry views—including suppliers—private sector input to this process will need to incorporate viewpoints and expertise outside of the industry in order to represent the broader national interest.

A final alternative would be an industry-government committee similar to the National Advisory Committee on Semiconductors (NACS), which was established by Congress with members appointed by the President.[14] NACS issued several reports over the years of its existence and disbanded in 1992. Although some experts credit NACS with helping to foster a closer industry-government partnership that has contributed to the resurgence of the U.S. semiconductor industry, others argue that its effectiveness was limited by political and other factors. Still, some useful insights can be drawn from this mixed experience. Clearly, a private sector advisory group cannot be fully effective in absence of government interest in its advice and willingness to incorporate that advice into policymaking.

From the preceding consideration of alternatives, several necessary characteristics for an effective new institutional mechanism can be identified: (1) it should have high-quality industry membership, but not be constituted or perceived as representing a "special interest"; (2) it should be a means to deliver regularized private sector input on policy questions of an interagency nature, preferably delivering that input to a high-level interagency group of officials; and (3) the effort to develop a shared vision for the aircraft industry must be supported by senior government and industry leadership.

In order to accomplish the task of consensus building and strategy implementation, the committee recommends the creation of a National Aviation Advisory Committee (NAAC) to report to an interagency group of responsible government officials. The primary responsiblity of the National Aviation Advisory Committee should be to create and further the implementation of a national vision for aerospace industrial development in the United States. Because of the interagency nature of this responsibility—reflected in many of the recommendations above—the committee suggests that this group report to the National Economic Council (NEC) or other appropriate group with interagency

---

[14]See National Advisory Committee on Semiconductors, *A National Strategy for Semiconductors* (Washington, D.C.: U.S. Government Printing Office, 1992).

responsibilities. The White House is already leading an interagency effort to reassess U.S. aeronautics and space policies, the effectiveness of which could be enhanced by regularized private sector input.[15] The NEC should take advantage of this existing effort in forming NAAC. As a channel for building private sector consensus on policy issues related to aircraft manufacturing, and by providing guidance for the wide variety of agency activities that affect the industry, NAAC would be a focal point for developing a shared vision and an effective strategy for the U.S. aircraft industry.

The National Aviation Advisory Committee would be composed of knowledgeable leaders from industry, academia, and elsewhere who could represent the national interest. Senior members of the government could attend meetings of this advisory committee in an ex officio capacity. To be effective, such a committee would need the full cooperation of the critical industrial sectors of the aircraft industry, including the lower-tier suppliers. NAAC could function well with a staff of two or three professionals detailed from industry or government agencies with responsibilities in aeronautics. The activities of NAAC could be reviewed periodically, and its agenda restructured as appropriate. The overall objective must be to maintain the leadership position of the U.S. aircraft (and its supplier) industry, and to maintain a strong domestic engineering and manufacturing base.

Besides developing a shared vision, other necessary tasks have been enumerated above: suggesting changes in tax and other policies to encourage capital investment and R&D by U.S. aircraft manufacturers, identifying critical technologies, developing guidelines for international technology transfer, and assessing international technology flows. As part of its mission, the National Aviation Advisory Committee should further the implementation of the other key recommendations made above, including new policies that promote rather than discourage civil-military integration, as well as greater commitment of resources and focus in government R&D programs on product-applicable aerospace technologies.

The National Aviation Advisory Committee should also be specifically charged with generating recommendations for policies to achieve balanced international flows of technology and symmetrical access. This task is central to continuing U.S. leadership in the global aerospace industry. In the past, Japan has utilized mandatory technology transfer to strengthen the technology base of its industries and enable companies to compete in global markets. Working with both U.S. primes and suppliers, NAAC should stimulate the development of new approaches—including incentives for transferring and utilizing technology from abroad—that advance the collective interests of the U.S. aircraft industry vis-à-vis the Japanese and other global industries.

A further important task is the removal of unnecessary barriers to cooperation between companies. In recent years, laws and regulations have moved in a

---

[15] "Washington Outlook," *Aviation Week & Space Technology*, September 27, 1993, p. 21.

positive direction on this front. This has led to encouraging developments in the aircraft industry—the partnership between U.S. companies and NASA on the High Speed Civil Transport program, and collaboration between GE and Pratt & Whitney in negotiations with the Japanese government on the HYPR program. U.S. antitrust laws and enforcement must continue to move toward a recognition that competition in many high-technology industries—particularly the aircraft industry—is global. Collaboration at the U.S. industry level should be permitted and extended to the supplier level in order to conserve resources in technology and program development, to respond quickly to global market needs with superior products, and—perhaps most important in the context of this report—to allow individual firms to work together when appropriate in bargaining with potential foreign partners so that they and the U.S. economy as a whole maximize the benefits of international collaboration.

Recommendations

- In order to implement the steps outlined here and provide an ongoing focus for strategy building for the U.S. aircraft industry, the committee recommends an independent National Aviation Advisory Committee be established by the National Economic Council.

Appendix A

The Importance of the U.S.
Aircraft Industry

The track record of the U.S. aircraft industry over the past fifty years constitutes one of the outstanding success stories of global competition. This
success and the importance of the aircraft industry to America's economic well-
being, national security, and technological leadership are testified to by numerous reports and experts.[1] The economic importance of the industry can be seen
clearly in the relevant statistics. The U.S. aerospace industry holds more than
half of the world market and ranks sixth among U.S. industries in total sales.[2]
In 1992, U.S. aircraft sales were $72.8 billion, and the combined trade surplus
for civil transports, engines, and parts was $23.7 billion.[3] Table A-1 contains a

---

[1]See National Research Council Aeronautics and Space Engineering Board, *Aeronautical
Technologies for the 21st Century*, (Washington, D.C.: National Academy Press, 1992), p. 1; Council
on Competitiveness, *Gaining New Ground: Technology Priorities for America's Future*, (Washington,
D.C.: Council on Competitiveness, March 1991), pp. 55-56; Michael L. Dertouzos, Richard K. Lester
and Robert M. Solow, *Made in America: Regaining the Productive Edge* (Cambridge, Mass.: MIT
Press, 1989) pp. 201-216; and U.S. Congress, Office of Technology Assessment (OTA), *Competing
Economies—America, Europe and the Pacific Rim* (Washington, D.C.: U.S. Government Printing
Office, 1991), pp. 341-358.
　　[2]The aerospace market is divided into several segments, including aircraft, missiles, space, and
related products and services. U.S. Department of Commerce, *U.S. Industrial Outlook 1993*
(Washington, D.C.: U.S. Government Printing Office, 1992), pp. 20/1-20/3.
　　[3]Aerospace Industries Association (AIA), "1992 Year-End Review and Forecast—An Analysis,"
December 1992. Note that AIA figures are somewhat different from the Department of Commerce
statistics appearing in Table A-2.

statistical comparison of the aircraft and aerospace industries with the chemical industry—another technology-intensive sector in which the United States is highly competitive globally. Table A-2 contains a breakdown of U.S. aircraft sales, Table A-3 lists aerospace export figures for 1992, Table A-4 indicates aerospace trade with Japan.

The aircraft and aerospace industries are also key components of America's larger technological enterprise. The aerospace industry accounts for about one-quarter of U.S. industrial R&D expenditures. Many of the technological competencies fundamental to competitiveness in transport aircraft diffuse to

TABLE A-1 1992 Industry Comparison—Aerospace and Chemicals

	Aerospace (aircraft)	Chemicals
Value of shipments	125.7 (54.0)	301.9
Share of gross domestic product (%)[a]	2.1 (1.0)	5.0
Employment[a]	695,000 (253,000)	853,000
Imports	12.7 (5.9)	25.1
Exports	42.2 (24.0)	44.2
Trade surplus	29.5 (18.1)	19.1
1989 R&D spending	20.3	11.5
1990 non-federally financed R&D spending	6.1	12.5
1990 non-federally financed R&D spending (% of sales)[a]	3.5	5.7

[a]Except for these items, all figures are current billion dollars.

SOURCE: U.S. Department of Commerce, *U.S. Industrial Outlook 1993* (Washington, D.C.: U.S. Government Printing Office, 1992); National Science Board, *Science and Engineering Indicators: 1991 Edition* (Washington, D.C.: U.S. Government Printing Office, 1991); and National Science Board, *The Competitive Strength of U.S. Industrial Science and Technology: Strategic Issues* (Washington, D.C.: U.S. Government Printing Office, August 1992).

TABLE A-2 U.S. Aerospace Exports (thousand 1992 dollars)

Product	Japan	Worldwide
New civil general aviation aircraft	13,381	580,799
New military aircraft	100,976	1,909,398
New civil heliopters	11,783	117,694
New civit passenger and cargo aircraft over 15,000 kg	2,574,413	22,378,686
Aerospace parts and equipment not elsewhere specified or included	1,089,140	10,146,951
Other civil and military aircraft, balloons, gliders	1,081	17,445
New and Used Civil and Military Piston Engines and Parts	1,777,348	315,734
New and Used Civil and Military Turbine Engines and Parts	449,172	636,220
Missiles, space vehicles, and parts	245,182	1,764,678
New and used civil and military aircraft engines and parts	466,519	6,683,953
New and used civil and military aircraft	2,703,859	26,419,249
Total	9,432,854	70,970,807

SOURCE: U.S. Department of Commerce.

other industries and contribute to the overall economy.[4] U.S. strength in the development and production of transport aircraft is also an important support for the defense industrial and technology base. Technology developed for commercial transports is often utilized in military programs; the production of commercial aircraft reduces military aircraft costs in companies that manufacture both; and commercial aircraft production helps to maintain the supplier and the work skill base in times of weak military demand.[5] Finally, the excellence of American-made aircraft has long played a major role in improving the safety and efficiency of the nation's air transportation system.

The aircraft industry—like many others—is regionally concentrated, so that its economic importance is felt unevenly throughout the country.[6] In

[4]These technologies include "system integration in the design and manufacture of complex, high-performance equipment; project management to meet demanding targets for performance, cost, and delivery; sophisticated manufacturing techniques for fabrication, testing, and assembly; and computer-integrated manufacture, factory automation, and large-scale integrated information processing" as well as "the more obvious ones that affect aircraft performance—aerodynamics, propulsion, advanced structures, and avionics and control . . ." National Research Council, *The Competitive Status of the U.S. Civil Aviation Manufacturing Industry* (Washington, D.C.: National Academy Press, 1985), p. 22.

[5]OTA, op. cit., p. 344.

[6]According to Boeing Commercial Airplane Group's brochure, "The Invisible Exporters," between 1987 and 1991 the Boeing Material Division procured an average of $10 billion per year in goods and services from suppliers in all 50 states. More than three-quarters of this amount was purchased from suppliers in four states: Ohio, California, Connecticut, and Washington. Of course, the larger first-tier suppliers in these states made purchases of their own, likely resulting in a greater geographic dispersion (including from overseas) at lower tiers.

TABLE A-3 U.S. Shipments of Aerospace Products (thousand dollars)

Product Description	1987	1988	1989	1990	1991
Aircraft	36,002,800	37,765,100	39,531,000	46,885,300	52,513,500
Military aircraft	16,862,300	15,044,400	14,832,900	14,108,700	15,622,000
Complete civil aircraft	12,491,743	16,019,855	17,421,046	24,864,289	29,780,098
Civil aircraft (fixed wing, powered)	12,145,669	15,453,662	17,108,080	24,608,896	29,550,713
Unladen weight not exceeding 2,000 kg	308,452				596,954
Unladen weight exceeding 2,000 kg but not exceeding 15,000 kg					802,657
Unladen weight exceeding 15,000 kg	11,837,217				28,151,102
Helicopters (rotary wing)	338,182	559,284	301,809	247,298	218,691
Other civil aircraft (nonpowered) and kits	7,892	6,909	11,157	8,095	10,694
Aircraft Engines and Engine Parts	18,821,900	18,866,700	19,903,900	21,580,200	21,314,900
Aircraft engines for military aircraft	4,205,600	3,214,200	3,342,000	3,265,800	2,967,600
Complete civil aircraft engines	2,841,150	3,753,689	4,358,246	5,335,475	5,778,444
Turbojet and turbofan	2,637,638		4,082,669	4,949,573	5,465,954
Turboshaft (turbo propeller):	203,512		275,577	385,902	312,490
Other, including auxiliary power units excluding missile and space engines	0	0	0	0	0
Aircraft Parts and Auxiliary Equipment Not Elsewhere Classified	19,528,900	20,545,400	21,294,500	23,081,800	25,288,200
Aircraft parts and auxiliary equipment, n.e.c.	15,817,800	16,331,000	18,155,900	19,618,100	22,155,100
Aircraft propellers and helicopter rotors	724,100	676,300	746,500	881,100	951,400

SOURCE: U.S. Department of Commerce.

TABLE A-4 1991 U.S.-Japan Trade in Aerospace Products (million dollars)

U.S. exports to Japan	3,907
Japanese exports to U.S.	661

SOURCE: Aerospace Industries Association, *Aerospace Facts and Figures 1992-1993* (Washington, D.C.: AIA, 1992), p. 122.

contrast to other high-technology sectors in which the globalization of markets and technological capabilities has prompted companies to multinationalize, aircraft manufacturers—at least at the level of airframe integrators and manufacturers of major subsystems such as engines and avionics—have generally *not* established their own offshore production and R&D sites.[7] The globalization of production and design has proceeded largely through international strategic alliances, consortia, and other types of supplier-partner relationships between nationally based companies.

Although U.S. companies continue to hold global leadership overall and in most important industry segments, the transport aircraft industry—including airframe integrators, engine makers, manufacturers of major avionic and structural components, and the broad supplier base—faces a number of significant challenges that threaten this leadership (see Table A-5). Global competition is intensifying—most notably in the large transport airframe market, where the Airbus Industrie consortium has leveraged significant support from four European governments to gain a large share of the market.[8] Also, as a result of declining defense budgets in the United States and elsewhere, fewer resources are available from military programs for R&D, training, and other investments—investments that have traditionally provided an indirect support to commercial product development. Further, the synergy between commercial and defense R&D has declined in recent years as military aircraft design increasingly emphasizes features such as stealth, high maneuverability, and short field landing capability. Finally, the global market for large commercial

---

[7] "The difficulty governments face in determining what constitutes a domestic firm, and therefore which companies are eligible for public support, is not a problem in this industry. There is little foreign direct investment in the aircraft business." George Eberstadt, "Government Support of the Large Commercial Aircraft Industries of Japan, Europe, and the United States," contractor document for the Office of Technology Assessment, May 1991, p. 11.

[8] See Gellman Research Associates, Inc., *An Economic and Financial Review of Airbus Industrie,* September 4, 1990. The European Airbus consortium members and their respective governments have argued that the indirect benefits that accrue to the U.S. aircraft industry from the defense budget are equivalent to the direct government support that Airbus members have received. The U.S. position is that these indirect benefits are not really equivalent and that, in any case, European aircraft makers also enjoy defense spillovers. Although a detailed treatment of the protracted U.S.-EC conflict over this issue is beyond the scope of this report, a number of the policy issues raised by the conflict and the 1992 U.S.-EC agreement are central to the committee's charge.

transports has experienced a significant downturn over the past several years due to sluggish demand for air travel. The impact of this cyclical slump through the aircraft supply chain has been exacerbated by structural problems afflicting the U.S. airline industry—traditionally the largest component of the aircraft industry's customer base.[9]

It is safe to assume that the aircraft industry will retain its economic importance into the next century, despite the current downturn in sales. The global market for air transportation and large transports is expected to grow significantly over the next several decades. Table A-6 shows that much of this growth is likely to occur in Asia. Further, in contrast to declining spillover benefits from defense to commercial markets, the importance of commercial transport manufacturing for maintenance of the defense industrial and technology base is likely to grow, both because fewer companies will be able to maintain extensive R&D operations on the basis of military work alone, and because increasing pressure for cost performance on the military side will require the incorporation of greater commercial discipline. The benefits that accrue to countries with a strong aircraft industry have always been compelling and have justified public policies of direct or indirect support in the United States and elsewhere. Europe, Japan, Russia, China, and other countries are pursuing a variety of policies to promote domestic aircraft manufacturing. The emerging environment for U.S. private and public policymakers is characterized by significant challenges, high stakes, and a complex field of players and interests.

---

[9]See testimony of Lawrence W. Clarkson, Corporate Vice President for Planning and International Development, The Boeing Company, and testimony of Thomas M. Culligan, Corporate Vice President, McDonnell Douglas, before the Subcommittee on Aviation, Committee on Public Works and Transportation, U.S. House of Representatives, on the "Financial Condition of the Airline Industry," Washington, D.C., February 24, 1993.

Table A-5 Aircraft Manufacturing Process and Supplier Structure

Materials	Structures	Integration	Delivery
Aluminum Composites	Fabrication Subassembly Tooling Machine tools	Engines Avionics Other components	Marketing Financing Certification

Materials	Structures	Engines	Avionics & Instruments	Other Components & Systems

Alcoa
Kobe Steel
Hercules
Toray
Yokahama Rubber
Union Carbide
Rohr
	U.S.
	  Vought
	  Grumman
	  Northrop
	  Rockwell International
	Japan
	  MHI
	  KHI
	  FHI
	  ShinMaywa
	  Japan Aircraft
	    Manufacturing
	Airbus
	  Deutsche Aerospace
	  Aerospatiale
	  British Aerospace
	  CASA
		Pratt & Whitney
		GE
		Rolls Royce
			Collins (Rockwell Intl)
			Allied Signal
			Honeywell
			Sundstrand
			Tokyo Aircraft Instruments
			Japan Aviation Electronics
			TRW
			Westinghouse
				Hamilton Standard
				Allied Signal
				Menasco
				Sundstrand
				Cleveland Pneumatic
				Shinko Electric
				Lear Siegler
				Kayaba

NOTE: The list of companies under each heading is included for illustrative purposes and is not an exhaustive list.
SOURCE: National Research Council Working Group on U.S.-Japan Technology Linkages in Transport Aircraft.

TABLE A-6 The Global Aircraft Market—Historical and Forecast Regional Shares of Average Annual Deliveries to Airlines (percent)

	1972-1981	1982-1992	1993-2000	2001-2010
United States	35	38	39	31
Europe	26	28	25	25
Asia-Pacific	20	24	27	33
Africa-Middle East	10	6	5	5
Latin America	5	2	3	4
Canada	3	2	2	2
Total market (billion 1993 dollars)	14.8	26.1	40.9	48.7

NOTE: Percentages may not total 100 due to rounding.

SOURCE: Compiled from data appearing in Boeing Commercial Airplane Group, *1993 Current Market Outlook*, March 1993, p. 3.5.

Appendix B

U.S.-Japan Technology Linkages In Airframes And Aircraft Systems

During the study process, the committee was briefed on U.S.-Japan cooperation and competition in various segments of the aircraft industry by industry and other experts. The material here and in Appendix C draws heavily on the insights of these experts and also incorporates information from published sources when this was available. Through this process, the committee was able to gain access to information on linkages that would otherwise be unavailable. On some points—particularly points of interpretation related to sensitive business issues—published sources of information do not exist. Readers should keep in mind that these accounts rely on individual expert viewpoints and interpretations.

BOEING COMMERCIAL TRANSPORT ALLIANCES WITH JAPAN

In the more than 20 years since the YS-11 program was canceled, Japanese activities in the airframe segment have been carried out mainly through alliances between the heavy industry manufacturers and Boeing. In the Boeing perspective, Japan is important as a market, collaborator, and potential competitor. As a wealthy island nation, Japan is a highly developed market for commercial airplanes and the largest foreign customer country for Boeing, even though the level of domestic air travel is low relative to the wealth of the country because of geography and the highly efficient rail system (see Figure B-1). Boeing projects that the total commercial jet market in Japan between 1993 and 2010 will be $60.5 billion in 1993 dollars (440 airplanes), second to the U.S. total of $280 billion and ahead of the rapidly growing Chinese market ($41 billion). Japan Air Lines (JAL) is the largest customer for Boeing's largest airplane, the 747 (having bought a total of 114); All Nippon Airways (ANA) is the largest foreign buyer for the 767 (having bought 82 thus far).

Boeing has procured parts and equipment from Mitsubishi Heavy Industries (MHI), Kawasaki Heavy Industries (KHI), and Fuji Heavy Industries (FHI) since the start of the 747 program in the late 1960s, with MHI and FHI supplying Boeing on the 757 and KHI on the 737. With the 767 program in the late 1970s and now the 777, the Boeing-Japan interaction has moved from one in which the Japanese companies "build parts to specification," to actual design and engineering interaction from the earliest stages of product development. Table B-1 lists the components built by the three "heavies" on various Boeing aircraft; while Table B-2 shows the involvement of other suppliers.

In looking at the U.S. versus foreign content of Boeing aircraft, on average U.S. content is 85 percent by dollar value across all models, and 60 to 70 percent of subcontracted work is given to U.S. firms. The big change over the past 20 years is the main fuselage sections. Northrop builds most of these parts on the 747, whereas most of the fuselage of the more recent 767 and new 777 models is built in Japan. However this has led to only a moderate shift in U.S. versus foreign content because the fuselage does not constitute a large percentage of the value of an airplane. Foreign content of the 777 will be 16.7 percent including engines (12.6 percent not including engines); foreign content is 12.2 percent for the 767 and 14.6 percent for the 757.[1] When describing their participation in Boeing programs, the Japanese companies use figures for percentage of the airframe by value, which are higher.

767 PROGRAM

Boeing had worked with the Japanese companies in the late 1960s when they supplied parts for the 747. Discussions concerning closer collaboration on future aircraft started in 1970; the 767 program was launched in 1978 and a contract was signed with the Japanese to supply parts. The first ship sets were delivered in early 1980. In 1991 the two sides renegotiated for a second 500

---

[1]Rolls Royce accounts for the largest share of foreign content. Boeing calculates these figures based on cumulative and projected engine purchases, and uses information provided by suppliers on foreign content of subsystems. It is generally possible to project engine purchases because airlines need to make a significant investment to support the maintenance of a particular engine. It is, therefore, very difficult (but not impossible) for engine makers to dislodge entrenched competition. For example, United generally buys Pratt & Whitney engines, and British Airways generally buys Rolls Royce.

sets. Boeing's primary motive in entering the alliance was the perception that Japanese participation would provide some market leverage. There were no formal offsets, laws, or requirements. Boeing negotiated a similar work share arrangement on the 767 with Italy's Aeritalia.

The 767 is a wide-body twinjet that can carry 260 passengers in a mixed-class configuration. In some versions, the range of the 767 exceeds 6,000 miles. The agreement was for a fixed-price purchase of the first 500 ship sets, which incorporated learning curve cost reductions over time. Boeing calculates Japanese work share as 15 percent of the airframe—this does not include airframe systems and constitutes about 6 to 7 percent of the total value of the airplane. This is a nonequity role. The Japanese have taken cost and market risks, and have covered their own tooling and other investments. The Japanese government provided funding through success-conditional loans for much of this investment. Boeing negotiated using its own production costs as a standard rather than bidding the work out competitively. Earlier procurements from the Japanese were competitively bid, as was some of the work the Japanese do on other models outside of these risk-sharing agreements. For example, FHI won the worldwide competition for the replacement of the 757 wing flap.

The first 500 ship sets were not guaranteed, which means that the Japanese consortium assumed the total risk for its work share. The price was fixed in dollars, which means that the exchange rate fluctuations during the 1980s played havoc with the planning of the Japanese companies. Since the yen appreciated overall, this has put cost pressures on the three heavies. Boeing believes that at this point the Japanese companies have made money on the 767 program overall, but results have varied greatly by year.

Boeing had already made the major design decisions when the 767 deal was signed with the Japanese. During the detailed engineering stage of the program, Japanese engineering personnel were stationed at Seattle for up to a year. Technology transfer between Boeing and its foreign partners was essentially limited by the hardware choices—Boeing did not give the Japanese (or the Italians) sensitive parts of the airframe. Engineering data exchange was conducted on a "need-to-know" basis. The Japanese were given engineering data necessary to design their parts through digital data transmission or magnetic tape. The Japanese were trained in computerized design techniques. Considerable transfer of component design technology occurred, but this constituted "old" technology from Boeing's standpoint. Transfer to the Japanese through program subcontracting probably allowed Boeing a higher return on this asset than alternate technology transfer mechanisms (such as licensing) would have, and the business arrangements were competitive.

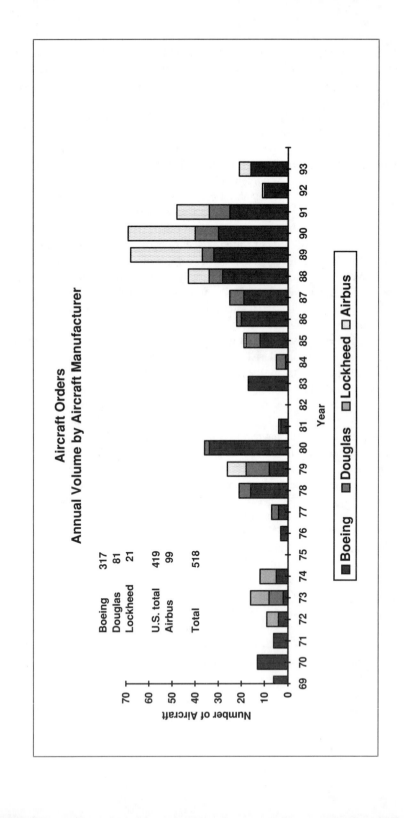

Aircraft Orders
Annual Volume by Aircraft Manufacturer

Boeing	317
Douglas	81
Lockheed	21
U.S. total	419
Airbus	99
Total	518

Number of Aircraft

Year

■ Boeing ■ Douglas □ Lockheed □ Airbus

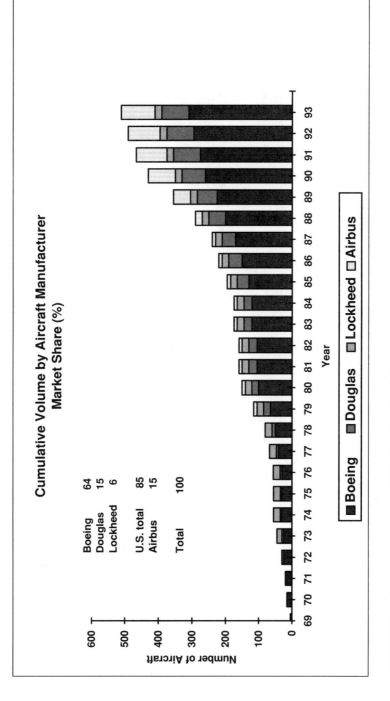

FIGURE B-1 Japan aircraft orders. SOURCE: GE Aircraft Engines.

TABLE B-1 Japanese Three Heavies: Boeing Involvement

Mitsubishi		Kawasaki		Fuji	
747	• Inboard trailing edge flap	747	• Outboard trailing edge flap	747	• Aileron • Spoiler
767	• Section 46 fuselage panel • Entry and cargo doors	767	• Sections 43 and 44 fuselage panels • Wing inspar ribs	767	• Wingbox fairing • Mlg. door
757	• Stringers	737	• Wing inspar ribs	757	• Outboard flaps
777	• Section 46 body panels • Passenger and bulk cargo doors • Section 48 tailcone	777	• Section 43 body panels • Section 44 body panels • Section 45 keel beam • Section 12 wing inspar ribs • Cargo doors	777	• Section 11 wingbox • Sections 11 and 45 integration • Section 49 wingbox fairing • Mlg. doors

SOURCE: Boeing.

TABLE B-2 Japanese Aerospace Industry Involvement in Boeing Programs

Manufacturing	Representative items and activities	Programs involved							
		707	727	737	747	757	767	717	777
Japan Aircraft Manufacturing Co.	Structural components					x	(x)		
ShinMaywa Industry Co.	Structural components					(x)	x		x
Japan Aircraft Manufacturing Co.	Galleys, lavatory modules		x			x	x		x
Teijin Seiki Co.	Actuators, servos			x	x	x	x	x	x
Shimadzu Corp.	Gearboxes, valves, actuators			x	x	x	x		(x)
Kayaba Industry Co.	Valves, actuators			x		x	x		
Yokahama Rubber	Lavatories, honeycomb core products			x	x	x	x	x	
Nippon Miniature Bearing	Bearings, motors				x	x	x		
Mitsubishi Electric Corp.	Valves, actuators				x	x	x		
Matsushita Electric Corp.	Entertainment systems, panel-mounted radio speakers, CRTs	x	x	x	x	x	x	x	
Sumitomo Precision Products Co.	Landing gear structures					(x)	(x)		
Koito Manufacturing Co.	Lights			x	x	x	x		
Tokyo Aircraft Instrument Co.	Instruments				x	x	x		
Sumitomo Electric Industries	Fiber optic couplers							x	
Suwa Seikosha Co.	Flat-panel displays							x	
NEC Corp.	Digital autonomous terminal access communications systems							x	

continued on next page

TABLE B-2 Japanese Aerospace Industry Involvement in Boeing Programs *Continued*

Manufacturing	Representative items and activities	Programs involved							
		707	727	737	747	757	767	7J7	777
Japan Aviation Electronics Industry	Flat-panel displays, air data intertial reference systems								
Koito Industry	Seats			(x)	(x)	(x)	(x)	x	
Toshiba Corp.	CRTs			x	x	x	x		
Showa Aircraft Industry Co.	Honeycomb			x	x				
Tenryu Industries Co.	Seats		x						
Shinko Electric Co.	Motors			x	x		x		
Shin Nippon Koki	Machine tools				(x)	(x)	(x)		
Kobe Steel	Titanium and steel forgings			x	x	x	x		
Furukawa Aluminum	Aluminum forgings, extrusions				x	x	x		
Daido Steel	Steel						x		
Toshiba Machine	Machine tools								
Sony Corp.	Entertainment systems								x

NOTE: (x) Significant indirect participation.

SOURCE: Boeing.

7J7/YXX

The next Boeing-Japan collaboration was on the 7J7 program, also known as the YXX in Japan. A memorandum of understanding (MOU) was signed in 1986 confirming Japanese participation as an equity partner in the development of a 150-seat short- to medium-range twinjet. This project would have constituted a significant increase in the Japanese role over the 767. The Japanese, through the Japan Aircraft Development Corporation, would have held 25 percent of the equity and would have been involved in all phases of design, production, and marketing. When the 1986 MOU was signed, the deal attracted considerable attention and some criticism.[2] Boeing argued that for each job created in Japan by the program, 23 would be created in the United States. Due to the subsequent shelving of the program, the collaboration returned to a lower profile. As in other collaborative international programs, U.S. government approvals—such as a Department of Commerce technical data export license—were needed for this project. The Departments of Commerce and Defense tend to take a restrictive stance on certain technologies such as those related to composites, but sensitivity is generally limited to design know-how rather than manufacturing processes.

Intensive discussions about market projections and other areas were initiated to start up the 7J7 relationship. A collaborative research program also commenced, with Boeing sharing some summary data from its generic subsonic research and the Japanese companies sharing data from the work they were doing with National Aerospace Lab and internally in fluid dynamics and the testing of new composite materials, flaps, and slats. Boeing sees high-speed aerodynamics as the fundamental technology to protect, and there has been no collaboration with the Japanese in high speed.

Although some support is still provided by the Japanese government, the 7J7 program has not yet been launched, and the short-term prospects are not particularly favorable. The market for the 150-seat aircraft has not coalesced. At the low end, it overlapped with Boeing's existing 737; at the high end, with the 757. Two technical developments outside of the Boeing-Japan negotiations have influenced this course of events. First, Boeing wanted to utilize an unducted fan turboprop engine, which it was working on with GE and which would deliver significant gains in fuel economy.[3] However, falling fuel prices in the mid and late 1980s made fuel economy a less critical concern for airlines. Secondly, contrary to earlier expectations, Boeing was able to extend the life of the 737 by fitting it with high bypass engines. The company had thought that

---

[2]Robert B. Reich, "A Faustian Bargain with the Japanese," *The New York Times*, April 6, 1986, Section 3, p. 2.

[3]David C. Mowery, *Alliance Politics and Economics: Multinational Joint Ventures in Commercial Aircraft*, (Cambridge, Mass.: Ballinger Publishing Company, 1987), p. 73.

the more advanced engine would not fit under the wing, but a solution was engineered, partly through computational fluid dynamics work at the National Aeronautics and Space Administration (NASA) Ames Research Center.[4] Because of the success of the extended 737 and prospective competition from the Airbus A320, a brand new airplane in the 150-seat class was much less compelling strategically for Boeing. More recently, Boeing has decided to develop an advanced version of the 737, pushing development of an all-new aircraft in that market segment further into the future.[5]

Both Boeing and the Japanese had put a significant amount of money into the program, and there is still a strong incentive for the Japanese not to let it formally die. The 7J7 continues to command a line item in the budget of the Ministry of International Trade and Industry (MITI). Low-level technical collaboration continues, but closer cooperation was delayed until the next major Boeing program, the 777.

777

Boeing, the three heavy industry companies, and the Japanese government worked out a "program partnership" for the 777 twinjet, which is due to enter airline service in 1995 and will seat 328 with a range of 5,000 miles in its initial version.

The structure of the deal itself is very similar to the 767, although Boeing originally offered equity participation similar to that contemplated for the 7J7. While the Japanese were interested in an equity share, Boeing set a minimum amount, and the Japanese were not prepared to assume a risk of that size. The 777 is a much larger airplane than the 7J7 was conceived to be, with correspondingly higher development costs. Development costs for an all-new jet such as the 777 are estimated to run about $5 billion.[6] As an equity partner, the Japanese heavies would be participating at a significantly higher level, and the business justification had to be compelling to their top managements. Apparently, the companies were not willing to assume that great a risk, despite some apparent pressure from the government to do so.

The Japanese ended up with nonequity participation and some increase in their role compared to the 767. There are differences between the two programs, the most obvious being increased Japanese work share. Boeing calculates that on the 777, the heavies are building 20 percent of the airframe,

[4]George Eberstadt, "Government support of the Large Commercial Aircraft Industries of Japan, Europe, and the United States," contractor document for the Office of Technology Assessment, May 1991, p. 74.
[5]As this report was being prepared, possible Boeing-Japan collaboration in the 737-X and YS-X programs was the subject of press reports. The Japanese government has supported research for a number of years on the 80 to100-seat YS-X transport, with the intention of eventually launching a program led by Japan with foreign participation. See Eiichiro Sekigawa, "Japan Mulls Joining 737-X Wing Project," *Aviation Week and Space Technology*, July 26, 1993, p. 32; and Jeff Cole, "Boeing May Aid Japan Suppliers in Building Jet," *Wall Street Journal*, September 8, 1993, p. A3.
[6]Jeremy Main, "Betting on the 21st Century Jet," *Fortune*, April 20, 1992, p. 102.

which is about 8 to 9 percent of the value of the airplane. The Japanese partners essentially build all of the fuselage parts except for the nose section. The increased share was not a negotiating issue—it was offered by Boeing. Also, rather than a contract for a fixed number of ship sets, the 777 agreement is in effect for the life of the program.

The Japanese are also more involved in designing the components that they are manufacturing. There were many more Japanese engineers involved in 777 development than in 767 development, with several hundred sent to Seattle during the most intensive design phase. Yet, as with the 767, the Japanese are limited in the engineering effort to their own work package. Structural testing—and the software and models needed to obtain results—are not shared, only the end results necessary for the Japanese to design the parts they will build. For example, Japanese engineers had access to the load data for the wing center section because they designed and will manufacture the wing box, but data on the outboard section were not made available. The most significant increases to the Japanese work share are the 777's wing box and the pressure bulkhead, both of which were designed and built by the heavies. Both are critical and somewhat tricky to manufacture, but they do not constitute "advanced technology" from Boeing's perspective.

The 777 is introducing several significant technical advances. The CATIA three-dimensional computer design system was used to make the design process as "paperless" as possible. CATIA was originally developed by Dassault Aviation of France, but Boeing has also made proprietary improvements. The full impact of CATIA will be known as the first aircraft are built and delivered. The process has not saved on the number of engineers needed to design the aircraft, but the hope is that it will cut down on the number of design modifications that have to be made after production begins. Boeing estimated that it would have 50 percent fewer modifications on the 777 than on the 767. As this is written, Boeing was running at the rate of 10 percent of 767 changes—a 90 percent improvement. The value of CATIA will be in the reduction of recurring costs by eliminating a significant percentage of the design anomalies that normally require correction during production of the first 30 or 40 ship sets. In addition to CATIA's role in streamlining the design process, the availability of design data in a digital form has enabled considerable manufacturing advances. This is discussed in more detail below.

Through a system of passwords, the access of Japanese engineers on-site and working at the computer system in Japan that Boeing set up for the 777 project is limited. The CATIA design software itself is "locked up," as is work on parts of the airplane unrelated to the Japanese work share. Attempts to get around the system would set off alarms. Japanese engineers went through a basic CATIA user's course, which takes about a week. Because the Japanese cannot access the software, it would not be possible for them to make improvements on CATIA in conjunction with the 777 program.

On-site, the visiting Japanese engineers were given access only to certain buildings, and sensitive manufacturing sites were accessible only with a Boeing escort. When two engineering teams get together, communication and information exchange naturally occur, most of which is beneficial to the project. How did Boeing control the people-to-people flow of technology? First, within Boeing, analysis and testing of the design are done by a separate group of engineers from those who work on design with the Japanese heavies. It is fairly straightforward to segment information on a need-to-know basis even within the company. Boeing provided a briefing to engineering and manufacturing personnel who would come in contact with the overseas partners, conveying the basic message that they should provide only what would be needed for the partners' work share. A management committee reviewed and decided on questions that arose in gray areas.

As in any program of this size, various management issues came up during the negotiations and design process. For example, after Boeing had determined how much work to give the heavies, the Japanese partners needed to reach agreement on dividing the work share. Apparently, this process was not completely straightforward. Also, once work got under way, the engineering resources of the heavies were stretched by the simultaneous demands of 777 and FS-X design work. Although the quality of the Japanese engineering effort was always outstanding, some adjustments were necessary during the design phase to keep the program on schedule. Also, from the Japanese perspective, the dispatch of large numbers of engineers to Boeing became quite expensive. The heavies would prefer to conduct as much of the interaction as possible within Japan in order to minimize travel and expatriate living expenses.

From a business and program development point of view, the partnership with Japan on the 777 has been very beneficial for Boeing. Since the product has not been introduced and its success in the market is not yet known, it is difficult to measure the bottom-line benefits, but order information thus far is promising. All three of the major Japanese airlines have ordered the 777.

Boeing's policy is to limit dependence on suppliers in the structures and airframe area (as opposed to the engine and avionics areas, where the breadth of technology is so great that the company has no choice but to be dependent). Boeing maintains the capability to manufacture all the components it buys from the Japanese. MITI has recently told Boeing that it wishes to encourage manufacturing technology transfer from Japan to the United States and has offered its assistance. Because of Boeing's relationship with the heavies, it is largely aware of what happens on the manufacturing side, and it reports no difficulties in obtaining access to technologies improved upon by Japanese partners. For example, one of the Japanese companies had developed a robotic skin polisher and responded positively to Boeing's request to license it. Another example is the method of laying up thick composite structures that Boeing learned from FHI in conjunction with the latter's work on the 767 main landing gear door. Boeing had been running structures of that size through the autoclave three times. FHI developed a method of laying up the composite material that required only two runs through the autoclave.

This open attitude might be different if Boeing were a potential competitor. For example, outside of the partnership with the three heavies, Boeing is contracting with Toray for composite materials to be used for the 777's tail. Boeing was interested in developing a second source, and Toray did license some of its technology to a U.S. firm, but in the end a competitive U.S. bid did not emerge. As a "next best" solution, Toray has built a facility in the Puget Sound region to supply Boeing.

Japanese Advanced Manufacturing Capabilities

During its study mission to Japan in June 1993, the committee had an opportunity to tour the manufacturing facilities of the Japanese heavies and several smaller aircraft suppliers. The committee was particularly impressed with the manufacturing capabilities of Japanese industry—much of it devoted to participation in the 777 and other Boeing programs. Here are selected examples of advanced aircraft manufacturing capabilities possessed by one or more of the heavies:

1. Fuselage panel drilling and riveting: In addition to its utility in the design process, CATIA allows for significant manufacturing process advances. By using the design data base to run manufacturing processes, much of the tooling that has traditionally been necessary for aircraft manufacturing can be eliminated. Using CATIA in conjunction with numerically-controlled machine tools to improve processes was inherent in the system from the beginning, but the Japanese have significant latitude in designing their own processes along with Japanese or foreign machine tool makers. Processes that the Japanese develop for the 777 must be approved by Boeing.

The Japanese have realized much of CATIA's potential in driving manufacturing in the fuselage panel drilling and riveting processes. Particularly impressive is the use of pogo sticks to support aluminum skins in drilling and riveting fuselage panels. The height and angle of the sticks as well as the hole locations are set according to the CATIA data base for particular panels. In addition to eliminating tooling, this reduces manufacturing cycle time and improves quality.

2. Preparing aluminum skins: Preparation of the aluminum skin for the panel requires chemical-milling of the panel around the window openings. Masking for the chemical-milling is prepared on a new large-scale numerically controlled five-axis, carbon dioxide laser that cuts a rubber mask laid on the panel. This machine also operates from the CATIA design data base. After chemical-milling, the skin is stretch formed to take the radius of the fuselage and is then trimmed and polished on another five-axis robotic machine.

3. Processes for thick aluminum parts: Although the aluminum skins used for many parts of the fuselage are thin enough to be shaped through chemical milling and stretch forming, some of the thicker parts would be made more susceptible to wear if this method were employed. Instead, thicker parts are shotpeened. The Japanese possess the advanced machinery necessary for this process and have acquired the complex know-how needed to use it.

All of the milling operations are done on large numerically controlled machines. To shape the curvature of the wing center section skin, a special machine generates the curved shape by shotpeening the skin in a controlled process while simultaneously shotpeening the surface for fatigue strength.

The machines used to mill the wing stringers are very high-speed, numerically controlled horizontal mills, about 15 feet long, that shape the stringers from a solid bar of aluminum. The wing spars are also milled on a universal five-axis numerically controlled machine.

4. Composites manufacturing: The Japanese heavies have made significant investments in composites manufacturing. Some of these are related to non-Boeing programs (such as the FS-X). Several Japanese companies possess the latest equipment to do immersion ultrasonic inspection of very large-scale composite aircraft structures. The equipment is also numerically controlled with automatic recording of inspection data, and is designed to detect subsurface flaws or lack of bonding in the composite structures. On the engineering side, the committee saw some excellent work being done on composite cloth configurations aimed at solving the fundamental problem of delamination in composite structures.

The overall impression is that various fundamental technologies have been distributed among the major players in the Japanese industry. From manufacturing processes involving fuselage structural components, to more highly loaded structures such as wing sections, to lightweight composite structures, which include moderately stressed composite landing gear doors as well as more highly loaded carbon fiber wing structures, Japanese aircraft manufacturing capabilities are state of the art.

The heavy investment in the most advanced robotic numerically controlled machines is clearly aimed at gaining a leadership position in high-quality, low-cost manufacturing. Although quality and manufacturing cost have always been a high priority in the U.S. aircraft industry, along with leadership in aerodynamics and systems integration, the committee gained a clear impression that the Japanese have placed a very high priority on winning in the arena of manufacturing quality while achieving cost leadership.

Boeing Manufacturing Capabilities

A subgroup of this committee also had an opportunity to visit several of Boeing's Washington State facilities that will manufacture the 777. During the past three years, Boeing has invested in excess of $2 billion in new factories, equipment, and office facilities aimed at achieving a quantum improvement in product quality and manufacturing productivity. This description of Boeing's capabilities is included to illustrate the scale of investment and types of advanced manufacturing technology currently required to stay competitive in the aircraft industry, to balance the discussion of Japanese capabilities, and to highlight the sorts of investments that most U.S. aircraft manufacturers are currently unable to make.

For example, the 777 assembly building in Everett includes all of the elements required for final aircraft assembly. The operation starts with buildup of spars and wing skins using the latest robotic riveting and bolting equipment. There are four huge fixtures for final assembly of left-hand and right-hand wing sections. Fuselage barrel sections are assembled from panels supplied by the Japanese heavies in huge "rollover" fixtures that permit access to assembly of the floor beams, with the floor assembled overhead and the barrel section rotated 180 degrees from its normal position. The floor beams are carbon fiber composite structures, the first such application of composites in Boeing commercial aircraft. The first 777 was rolled out on April 9, 1994, with plans to commence flight tests in June 1994.

The Auburn sheet metal shop is another new facility in which up to 10,000 different structural components, from simple brackets to the huge hydroformed beams that connect the wings through the wing box, are manufactured. Facilities include very large, new, horizontal milling machines for cutting multiple elements of complex geometric shapes. The machining center is computer driven from dispatch of raw material through delivery of finished parts. The machining instructions are contained on compact discs that are inserted in the machine by the operator. The plant contains some of the world's largest hydroforming and stamping equipment.

These investments will likely enable Boeing to achieve major improvements in cycle time. Previously it took an average of 40 days to process a part from order input to product output. Today it takes about nine days, the objective being a five-day cycle. Current efforts are focused on reducing product variability by using techniques such as statistical process control.

The Fredrickson wing spar and skin mill facility is also new. In this plant are four huge wing skin milling machines with vacuum milling beds up to 210 feet in length. Each machine is capable of milling two wing skins simultaneously. In addition, there are similar special milling machines for machining the wing spars. The plant includes special facilities for shotpeening the spars and edges of the wing skins, automated anodizing, and painting. The plant delivers the complete wing skins and spars to the Everett facility, where the wing skin-spar assembly is completed.

Boeing has also made significant investments in composites manufacturing capability at Fredrickson. The facility includes four large-scale tape lay-up machines, with the entire process carried out in an atmospherically controlled (positive-pressure) building. Two new 40-foot-diameter autoclaves, with front and rear door loading, are operational. All trimming and cutting operations are done by a computer-controlled water jet cutter. The compound curved structures are supported on a pogo stick bed driven by the CATIA data base.

Impacts

Thus far, Boeing's relationships with Japan have been quite beneficial to it in a business and strategic sense. Boeing's basic philosophy is that Japan will be a major player in aircraft, and that it is preferable for the major firms to be teamed with Boeing rather than allied with one of Boeing's rivals or mounting an independent challenge. The Japanese have not collaborated in a significant way with either Airbus or McDonnell Douglas on commercial transports, and have not become an independent force thus far. Airbus has been actively looking for Japanese participation in its programs. The Eurpean consortium has sold A320s and A340s to ANA, with ANA obtaining important European landing rights at about the same time. Kawasaki has one contract for the A321, which is the first supply contract between one of the heavies and an Airbus partner.

Boeing has received high-quality components delivered on time at a price that U.S. suppliers would be very hard pressed to beat. The risks assumed by the Japanese (in the form of success-conditional loans by the government and the companies' own investments) have allowed Boeing to avoid the high financial leveraging necessary for earlier projects like the 747. The Boeing relationship has provided the Japanese heavies with a relatively low-cost, low-risk means of entering the global airframe field. Participation in Boeing programs—particularly the 777—has allowed the Japanese heavies to implement advanced manufacturing techniques in producing modern technology aircraft, but they have not obtained Boeing's most critical technologies.

Perhaps the most critical technology in design is knowing how to make the end product do what it is supposed to do on paper. This is a very difficult process, one that even established players find daunting. Boeing's track record is quite strong in this area. Because the engines are a critical determinant of performance, Boeing audits the engine makers to assess whether new products are likely to meet targeted performance specifications and then estimates the size of any shortfall. This engine audit process is a part of Boeing's organizational knowledge base. Another closely held management technology is the know-how needed to guide a program through the safety certification process and to interact with the Federal Aviation Administration and air safety agencies of other governments.

Up to this point, the Japanese have been content to continue in the role of risk-sharing supplier. The heavies will likely continue to receive government support for Boeing projects as long as they can show that they are receiving increased work shares with greater technical sophistication. Aerospace is a significant but not overwhelming share of the overall business of the heavies. Defense and commercial aircraft programs must compete for resources with other divisions, and the road to the chairmanship of MHI, KHI, or FHI has not traditionally led through the aerospace division. The companies have not significantly "grown" their aerospace activities—there are perhaps 2,000 to

3,000 aerospace engineers in Japan, whereas Boeing employed 9,000 on the 777 program alone.

Although Japanese defense budget cuts will likely increase industry's appetite for commercial work, Japan also faces some constraints as it reassesses its long-term strategy. The Japanese heavies have failed twice in independent programs, and it is Boeing's policy not to participate in a program at less than 50 percent equity. Further, significant participation in McDonnell Douglas commercial programs might be more costly and risky than continuing with Boeing, and collaboration with Airbus is problematic because the heavies would presumably need to take work share away from the Airbus members themselves. In the case of Boeing, the Japanese are largely building components that Boeing would have contracted out anyway.

Although it is important to recognize the constraints currently facing the Japanese aircraft industry, there is still no question that Japan has built a considerable aircraft technology and business base over the past several decades. Significant changes in the global environment, including the emergence of the Russian industry and other new players, may present Japan's aircraft industry with opportunities to move beyond existing constraints. Japanese capabilities, particularly in manufacturing, will allow its industry to continue expanding its global role into the next century.

MCDONNELL DOUGLAS

Commercial Programs

McDonnell Douglas's involvement in Japan stretches back over 40 years. JAL has operated a variety of Douglas products (DC-3, DC-4, DC-6, DC-7, DC-8, DC-10, and MD-11) since 1951. Japan Air Systems (JAS), the major domestic carrier along with ANA, is also a longtime Douglas customer. In addition, the trading company Mitsui & Co. played a major role in financing the launch of the MD-11 program. Yet in contrast to growing involvement by Japanese airframe manufacturers in Boeing programs over the past two decades, Japanese firms have remained subcontractors in McDonnell Douglas commercial programs.

Still, even this limited involvement has led to growing Japanese capability in a number of structures and components areas, particularly composites. In the early 1970s, MHI won a contract to supply the metallic tail cone for the DC-10, and is now manufacturing a composite tail cone for the MD-11. Also, FHI supplies a composite outboard aileron for the MD-11, which meets the targeted weight at a cost equivalent to aluminum. Table B-3 shows the Japanese suppliers for the MD-11 program.

McDonnell Douglas has had several other collaborative relationships with Japanese companies and the Japanese government over the years in aerospace fields such as satellite launch vehicles and helicopters. However, the interaction

TABLE B-3 Japanese Suppliers on the MD-11 Program

Manufacturer	Product
Fuji Heavy Industries	Outboard aileron
Nippon Hikoki	Underwing barrel
ShinMaywa (through Rohr Industries)	Wing/tail pylon
Yokohama Rubber	Portable water tank
Teijin Seiki	Elevator activator
	Sleet activator

SOURCE: McDonnell Douglas.

in fighter aircraft is the one most relevant to this study and constitutes a good starting point for a discussion of U.S.-Japan collaboration in military aircraft.

Military Programs: F-15 Licensed Production

U.S.-Japan licensed production of the F-15 was an important step in the evolution of U.S.-Japan collaborative military programs. As noted earlier, Japanese companies had assembled the North American F-86 in the 1950s, and moved on to the licensed production of the more advanced Lockheed F-104 in the 1960s, and the McDonnell Douglas F-4 in the 1970s. In the mid-1970s, Japan began to consider options for replacing the older fighters in the Air Self Defense Force (ASDF) arsenal. The F-15 was chosen over several rivals mainly because of its weaponry, radar, and other aspects of its technological sophistication as an "air superiority" fighter. This decision and the subsequent licensed production agreement were reached relatively soon after the fighter was first deployed in the United States.

There were early security concerns in the U.S. Defense Department over the transfer of advanced technology through F-15 licensed production. Japan is still the only U.S. ally that has been allowed to produce the aircraft. Concerns about the economic and competitive implications of F-15 technology transfers were raised only after the program was launched.[7] In initially deciding to go forward, the broad strategic and political rationale for Japanese production—primarily a greater contribution to regional security from a more militarily capable Japan—prevailed without a great deal of contention in the U.S. government.

---

[7]In the original 1978 MOU, there were no provisions for the "flowback" at no charge of technological improvements made by the Japanese. An amended 1984 MOU did contain explicit provisions. See Michael W. Chinworth, *Inside Japan's Defense: Technology, Economics and Strategy* [Washington, D.C.: Brassey's (U.S.), 1992], pp. 109-110.

The United States was committed to provide technologies and data necessary for Japanese production of the F-15, with the exception of items such as design data, radar, electronic countermeasures, software, and source codes, which were classified as "nonreleasable." The extent of this "black boxing" was greater than in the F-4 program and, according to some experts, provided a motivation for Japanese industry to pursue the independent Japanese development of the country's next fighter in the mid-1980s. Still, the technology transfer was substantial in terms of quantity, and it has been argued that the level of technology transferred through F-15 licensed production was higher than in previous bilateral programs.[8] Table B-4 shows the technologies transferred to Japan by McDonnell Douglas in the F-4 and F-15 programs.

Much of the technology transfer connected with the F-15 program has taken place through commercial licensing and technical assistance agreements between individual companies. Although these agreements are subject to U.S. government export approval, Department of Defense (DOD) program officers and even McDonnell Douglas are not equipped to stay fully abreast of technology transfers at the subcontractor level. At the government level, Japanese industry and government have continued to request technical information connected with the F-15, including releasability requests for technologies that the U.S. had provided in black boxes. There was sometimes disagreement among DOD management over these requests, with the F-15 system program office inclined to urge denial and higher levels tending to approve.

It was often difficult to balance Japan's justifications with concerns about protecting U.S. design know-how. Economic concerns about the potential for F-15 technology aiding Japan's commercial aircraft capabilities gained credence as the program progressed. Japan generally justified requests by claiming that release of a given technology would speed production schedules, reduce maintenance times, alleviate parts shortages, and reduce the costs of maintaining large inventories of spares. Some of these requests were understandable—a number of the U.S.-made components had high failure rates, with repair sometimes requiring shipment back to the United States. Japanese companies also reported cases in which American supplier counterparts either lost orders for spare parts or filled duplicate orders. This materially affected the operations of Japan's deployed F-15s and provided an impetus for independent development of the FS-X.

Still, requests for technical information, Japanese delegations, and other mechanisms were often used in attempts to gain information that was only indirectly connected with Japan's capability to produce and maintain the aircraft. When consideration of the next-generation fighter began in the mid-1980s, DOD officials were also forced to consider whether Japanese requests

---

[8]"The initial list of technical data to be made available to the Japanese in the F-15 program, for example, consisted of 21 pages listing more than 300 items that in turn consisted of everything from single drawings and rolls of microfilm to magnetic tapes and boxes of microfiche." Ibid., p. 117.

TABLE B-4 Military Aircraft Contribution

Technology Transfer

F-4/F-15 License and Technology Assistance Agreements (LTAA)
    Technical data (excluding design data)
    Technical assistance
    Factory training
    Tooling
    Production test equipment
    Mobile training unit
    Knockdown assemblies
    Follow-on material

F-4 Technologies
    Titanium machining
    Titanium forming
    Wire bundle manufacturing
    Stability augmentation and flight control system integration

F-15 Technologies
    Boron and graphite composite
    Titanium tubing
    Digital multiplex bus system integration
    Limited software development capability
    Fly-by-wire flight control integration

NOTE: No design technology or design data has been transferred.

SOURCE: McDonnell Douglas.

were really motivated by a desire for technology that could aid the development of an indigenous fighter. In some cases, the competitive implications were felt more quickly. Soon after Japan Aviation Electronics (JAE) was licensed to produce Honeywell's ring laser gyro-inertial navigation unit, it began marketing a similar system.[9] In the case of the AP-1 mission computer manufactured by IBM, the American company observed the Technology Research and Development Institute (TRDI) and Japanese corporate R&D programs targeted at developing a domestic mission computer for the FS-X, and decided not to contribute to these efforts by licensing its technology. The Japanese programs proved successful anyway—the FS-X mission computer will be indigenous. These two cases illustrate the difficulties faced by U.S. companies in making licensing decisions in areas where Japanese companies are capable and where government and industry are determined to reduce dependence on foreign suppliers. The potential for short-term licensing income, the competitive implications of technology transfer and other factors must be carefully balanced.

[9]Ibid., p. 121.

In assessing the significance of F-15 licensed production for the development and technological capability of Japan's aircraft industry, analysts present a mixed picture. On the commercial side, a large number of Japanese suppliers make similar components for the F-15 and for the Boeing 777.[10] However, many of these Japanese suppliers were already making similar components for Boeing prior to the launch of the F-15 program. The F-15 work was beneficial in enabling Japanese suppliers to invest in new equipment more rapidly, to make incremental improvements in technology, and to cross-fertilize capabilities from military to commercial work and from aircraft manufacturing to other businesses. This process was aided by the close integration of Japan's military and civilian industrial bases in aircraft.[11] The disagreement among analysts centers on the ultimate significance of F-15 technology transfers for commercial aircraft competitiveness, as distinct from the benefits presented by the work itself.

The impact is somewhat clearer on the military side. There is general agreement that the F-15 experience lifted the confidence of Japan's aircraft industry and that Japanese companies receiving technology through F-15 production were in a better position to supply the subsequent FS-X program. The denial of U.S. technology also had an impact. The black boxes provided a focus for TRDI and industry R&D efforts and motivated Japanese industry to pursue an indigenous FS-X. Still, the difficulties that have been widely reported in connection with the development of the FS-X show that the Japanese did not gain the capability to independently design and develop an advanced fighter through F-15 licensed production. Although the experience lifted the confidence of Japanese industry to perhaps unjustifiable levels, subsequent developments have exposed continuing weakness in certain key areas.

FS-X

Sweeping conclusions about the FS-X are premature since the development phase is only now reaching a conclusion, and critical issues such as the actual performance and procurement of the aircraft have yet to be resolved. However, it is safe to say that the process of structuring this Japan-U.S. codevelopment program marked something of a watershed in Japan's security policies and U.S.-Japan relations.

Soon after the launch of F-15 licensed production, the Japan Defense Agency (JDA), the Air Self Defense Force, and industry began considering options for replacing the domestically developed F-1 fighter. Although the F-1

---

[10]U.S. General Accounting Office, "Technology Transfer: Japanese Firms Involved in F-15 Coproduction and Civil Aircraft Programs," GAO/NSIAD-92-178, June 1992.
[11]David B. Friedman and Richard J. Samuels, "How to Succeed Without Really Flying: The Japanese Aircraft Industry and Japan's Technology Ideology," in J. Frankel and M. Kahler, eds., *Regionalism and Rivalry: Japan and the U.S. in Pacific Asia* (University of Chicago Press, 1993), pp. 43-47.

only entered service in 1977, its limitations made it "useless for any real combat from the day it was deployed."[12]

Even prior to some of the negative F-15 experiences with repairs and spare parts, Japanese industry and some elements in the government began the process with a presumption in favor of a domestically developed fighter. Increasing domestic content, gaining greater managerial control over the program than was possible in a coproduction arrangement, and controlling costs (costs of licensed U.S. aircraft increased by an average factor of four with each program from the F-86 to the F-15) were all considerations. Perhaps the most important factor was an underlying sense that Japan's position in the aircraft industry was fragile and that passing up domestic development would consign Japan to a follower role forever.[13] However, some Japanese policymakers were more cautious. Even at the early stage—before U.S.-Japan relations became a major factor in the decision—some MITI officials worried about industry overreaching. There was also a general recognition that even an "indigenous" fighter would require significant foreign inputs and technology (engines, systems integration).

Although the process of considering options began in the early 1980s, the U.S. government did not involve itself very extensively. By the time serious feasibility studies were launched in 1986, the momentum in Japan for a domestic aircraft was quite strong. The JDA set specifications that could not be met by existing aircraft, and MHI completed preliminary designs for a domestic aircraft—with an unrealistically low estimate of development costs.[14] During 1986, DOD became increasingly concerned with the specifications and low development cost estimates, and began a more aggressive push for the FS-X to be based on an existing U.S. design. The McDonnell Douglas F-18 and the General Dynamics F-16 were the leading candidates. DOD's report in 1987 that the cost of a new Japanese design would be two or three times higher than MHI and JDA estimates gave support to Japanese opponents to the indigenous option in the Ministry of Foreign Affairs and elsewhere. In October 1987, after a heated struggle within the Japanese bureaucracy and in the wake of the Toshiba Machine "incident," the United States and Japan reached an agreement to "codevelop" an FS-X based on the F-16 design.

From the start, the two countries conceived codevelopment differently, making it an attractive political solution but ensuring problems later. The Japanese assumed that a Japanese company would manage the process of developing an indigenous aircraft, with selected foreign technologies incorporated as necessary. The U.S. conceived the joint improvement of an existing aircraft, with a priority on ensuring "flowback" of Japanese technology based on know-how transferred by the United States.

[12]Chinworth, op. cit., p. 135.
[13]Ibid., p. 138.
[14]Ibid., p. 143.

Through late 1987 and 1988, an MOU for the development program was negotiated. DOD aimed for a 40 percent U.S. development work share (excluding work on the engines), but this posed problems for the Japanese because more than half of the development costs were slated to go toward domestically developed avionics. A U.S. share of 40 percent would mean that there would be very little development work left for Japanese companies in areas such as composite wing technology. The MOU was finally signed in late 1988.

With the Bush administration coming into office in early 1989, congressional concerns over the FS-X agreement were raised in confirmation and other hearings. Contentious debate over the agreement continued through the spring of 1989, with opponents arguing that F-16 technology transfers would contribute to Japanese competitiveness in commercial and military aircraft, to the long-term detriment of U.S. industry; that "off-the-shelf" Japanese procurement of F-16s would cut the huge U.S. trade deficit with Japan while addressing Japan's security needs more economically; and that Japanese technical capabilities were not high enough for the flowback provisions to deliver many benefits to the United States. U.S. proponents argued that significant U.S. participation in the FS-X program was better than none at all, that Japanese procurement of unmodified F-16s was not a realistic scenario, and that flowback would bring considerable benefits.

In the end, congressional opponents were not able to stop the FS-X agreement, but they were able to force DOD to gain a "clarification" of several key points. First, the Japanese explicitly committed to a 40 percent U.S. work share during the development phase and to providing access to Japanese-developed technologies. Second, the denial of several key F-16 technologies—including computer source codes, software for the fly-by-wire flight control system, and other avionics software—was made explicit. The Japanese had perhaps been counting on getting this technology, but DOD had never allowed technology transfer in these areas before—to Japan or any other country.

The clarification exercise probably had little material impact on what would actually transpire during the development phase, but it did serve to illustrate that U.S. policy toward defense technology collaboration with Japan could no longer be made without considering the economic impacts. The episode threw into sharp relief the contrast between the contentious divisions over Japan policy in the United States and the much more united front—albeit with some bureaucratic infighting—that Japan presents to the United States in bilateral negotiations. In addition, the contention left heightened resentment on both sides. Many Japanese opinion leaders, in particular, resent codevelopment as having been forced on Japan by the United States.

The development phase is now nearing completion, and first flight is projected for September 1995. Development was delayed during 1991 and 1992—in part because of sanctions placed on JAE after it was found to have violated export controls. Some observers expect that the development phase will be termed a "success," but the prospects for actual procurement are still uncertain.

Consideration of the U.S.-Japan MOU on FS-X production was slated to begin in 1994. One complication is possible disagreement over development issues, particularly flowback. The original development MOU defined four areas of nonderived technology, meaning that U.S. companies could license technologies in those areas for a fee, but would be entitled to Japanese developments in other areas at no charge.[15] Some observers believe that this designation was arbitrary and made subsequent Japanese requests to reclassify other technologies inevitable. In early 1993, news reports indicated that JDA was indeed demanding the reclassification of fifty technologies.[16] Although the FS-X is politically dormant as this is written, controversy could be reignited over the issue of derived versus nonderived technologies or over the production MOU.

By keeping in mind the considerable remaining uncertainties, it is possible to identify some key questions concerning the implications of the FS-X as a U.S.-Japan technology linkage and to catalogue areas in which analysts generally agree or disagree. The three key issues are as follows: (1) What are Japanese aircraft capabilities as illustrated by the FS-X? (2) What will be the impact of technology transfers from the United States to Japan? (3) What is the value of technology transfers from Japan to the United States?

On the first point, it is already evident that the FS-X will not be the "superplane" that the Japanese originally claimed it would be. Some of the technologies that Japan was originally planning to incorporate (canards) did not perform as well as expected and have been removed from the design. Despite some attempts to blame the U.S. side for cost and schedule problems, there is no question that the original Japanese projections of FS-X capabilities were unrealistic and that the hubris evident in the late 1980s has been deflated to some extent.[17]

The long-term implications of United States to Japan technology transfer are still unclear. Although the source codes and other critical items listed above were not transferred, the considerable modification of the F-16 necessitated the transfer of design and systems integration technology from the United States to Japan—a first in bilateral military programs. Although much of this technology is "old," analysts have pointed out that Japan has developed competitive

---

[15]The four nonderived technologies are all in avionics: the phased array radar, the inertial reference system, the integrated electronic warfare system, and the mission computer. The composite wing is considered derived.

[16]"Bei ni 'Dokuji Gijutsu' Nintei Yokyu" (Demand to U.S. for "Independent Technology" Designation), *Nihon Keizai Shimbun*, February 23, 1992, p. 1.

[17]Two representative pieces from that period are "Nihon no yui gijutsu, Beikoku no yui gijutsu," *The 21*, July 1989, pp. 28-29, and the occasional "Militeku" (Militech) series that ran in the *Asahi Shimbun* from January 30, 1989 to March 4, 1989.

capabilities starting from old know-how.[18] While it appears that the level of technology transfer to Japan that is occurring through the FS-X program is higher than what occurred in connection with the F-15, the extent to which the Japanese will be able to capitalize on it—in military as well as commercial aircraft development—is still an open question.

Finally, there is also considerable disagreement about the value of Japanese technology developed for the program that U.S. industry will have access to (either as flowback or through licensing). Observers disagree on the quality of Japan's phased array radar technology. While General Dynamics is reported to have found the flowback of composite wing technology from Mitsubishi to be useful, with the sale of the fighter division to Lockheed—which has been viewed as superior to General Dynamics in composites technology—the ultimate value of technology transfer in this area is uncertain. It is safe to say that the value of the technology flow to the United States is nowhere near the value that has flowed to Japan through this program.

---

[18]Chinworth, op. cit., p. 155. He also remarks on the irony of the pains taken by the United States to avoid transferring design technology during the F-15 program, only to transfer F-16 design technology to the same companies a few years later.

Appendix C

U.S.-Japan Technology Linkages In Aeroengines [1]

Because jet propulsion is the key enabling technology underlying commercial and military aviation as we know it today, the engine industry plays a special role in the aircraft supplier base. U.S.-Japan technology linkages in the engine business are extensive and long-standing, and they cover a range of mechanisms. The global context of growing international alliances in the commercial and military jet engine businesses is also important. The experiences of the two American engine makers—General Electric and Pratt & Whitney—have been somewhat different.

GE AIRCRAFT ENGINES

As a corporation, General Electric has a 90-year history of involvement with Japan. GE Aircraft Engines has been involved with Japan for more than 40 years (see Figure C-1). GE was involved with the first Japanese postwar military aircraft program starting in 1953 with the J47 engine for the Japanese version of the F-86 fighter. Over the next several decades, GE's J79 engine was chosen to power the Japanese versions of the F-104 and F-4. GE's relationships with Japan during this period involved sending kits to Ishikawajima-Harima Heavy Industries (IHI) for assembly and testing, with some components manu-

---

[1]Appendix C, like Appendix B, relies on the insights and interpretations of individual experts.

factured by IHI. Collaboration took similar forms on several military turboprop and helicopter programs. GE also has a long-standing relationship with IHI in aero-derivative marine and industrial engines. The IHI connection has provided GE entree to the Japanese Marine Self Defense Force (MSDF), helping it to fundamentally displace Rolls Royce over the years.

As for activities in commercial jet engines, it is important to remember that GE did not emerge as a true force in the commercial business until the 1970s. GE's first sales to Japan were to Japan Air Lines (JAL) in the mid-1960s, with the CJ805 engine on the Convair 880. This engine was a derivative of the J79, had a number of in-service problems, and did not live up to its technical expectations. At that point, GE exited the commercial market for a time, reentering in 1971 with the next generation of high-bypass technology with the CF6-6 and CF6-50 engines for the DC10-10 and the DC10-30. This was followed by the introduction of the CF6-50 engine on the 747 and the Airbus A300 in 1973. GE learned several lessons that it put to work over the next several decades. As a result of the CJ805 experience, GE built an excellent customer support organization. Specific to Japan, GE learned that it is important to completely fulfill the expectations of Japanese customers. GE did not make another commercial sale in Japan until it reentered the commercial engine business in the late 1970s and did not make a sale to JAL until the mid-1980s, when JAL selected the CF6-80C2 for their 747-400s.

The opportunity to reenter Japan came when All Nippon Airways (ANA) decided to upgrade and expand its fleet with the latest generation of wide-body aircraft. The initial opportunity with ANA led to a tremendous fleet of follow-on sales for 747s, 767s, and A320s. Japan Air Systems is also a major customer (see Figure C-2). The big competitive issue today involves engine selection for the 777s that JAL has already ordered. As the Japanese airlines have expanded their fleets to accommodate more traffic growth, GE's market share has increased. This has recently been augmented by the sale of CFM56-powered Airbus aircraft in Japan. One interesting characteristic of the Japanese airlines is that they generally do not want to be the first to buy a major new aircraft or engine. They desire the company of at least one other major airline to ensure that the needed support will be available if there is a problem. The manufacturer's product support infrastructure is a major consideration in the selection of the engine.

GE has focused its engine collaboration in Japan with IHI. The major collaborative programs relevant to this study are the GE90 and the F110 engine for the FS-X. In addition to GE, IHI collaborates with Pratt & Whitney, Rolls Royce, and others. This contrasts to GE's European partner, France's Snecma, which has limited itself to GE. GE does not consider this a problem, because IHI has not involved itself in technical development programs for competitive engines, even though its involvement with programs such as the PW4000 or the Rolls Royce Trent may be large in terms of manufacturing work share. Further, GE's collaboration with IHI in developing a commercial engine is fairly recent,

FIGURE C-1 GE Aircraft Engines Relationships

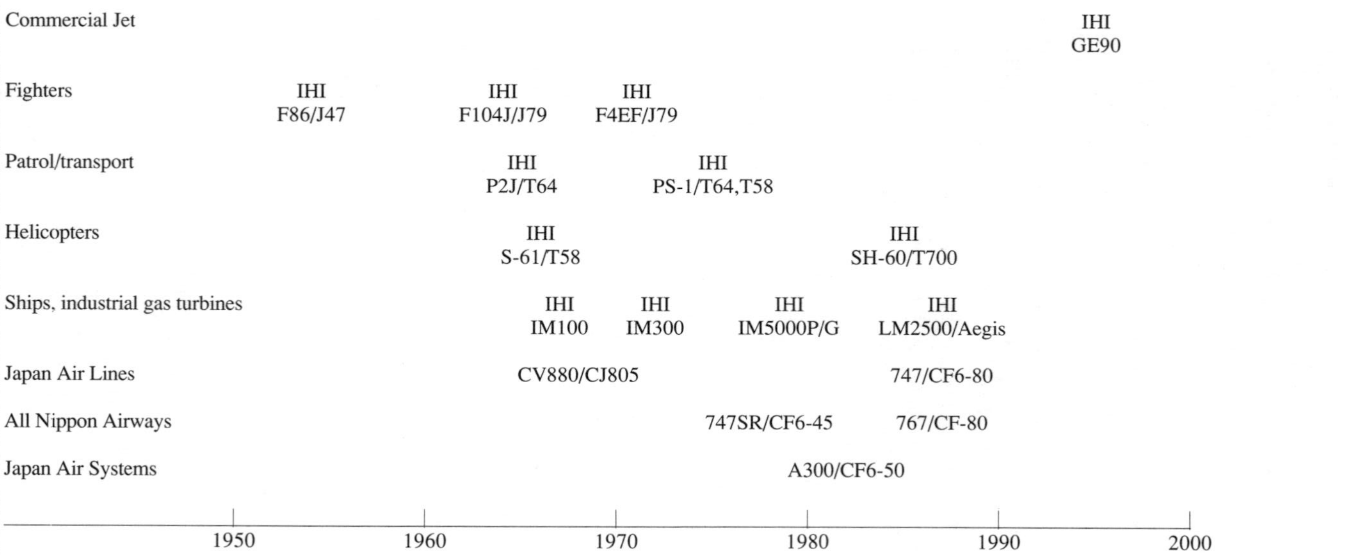

	1950	1960	1970	1980	1990	2000
Commercial Jet						IHI GE90
Fighters	IHI F86/J47	IHI F104J/J79	IHI F4EF/J79			
Patrol/transport		IHI P2J/T64	IHI PS-1/T64,T58			
Helicopters		IHI S-61/T58		IHI SH-60/T700		
Ships, industrial gas turbines			IHI IM100	IHI IM300	IHI IM5000P/G	IHI LM2500/Aegis
Japan Air Lines			CV880/CJ805		747/CF6-80	
All Nippon Airways				747SR/CF6-45	767/CF-80	
Japan Air Systems				A300/CF6-50		

SOURCE: GE Aircraft Engines.

having only begun with the GE90. As for possible GE partnering with other Japanese engine companies, a good opportunity for collaborating with Mitsubishi Heavy Industries (MHI) or Kawasaki Heavy Industries (KHI) has not presented itself, and GE has not felt compelled to seek one out.

GE90

The GE90 is the first of what GE hopes will be a new family of large engines to power the next generation of commercial transports. In the late 1980s, GE determined that the derivative path of the CF6 family had served its purpose, and decided on developing a new family of engines based on proven technology. This program was centered around the thrust requirements of Boeing's 777 family of aircraft. The major question was how the program would be structured. In order to spread risk, obtain maximum leverage of development resources, and gain global market opportunities, it was decided that the program would be structured around GE's existing international relationships. Snecma, the French engine maker that is GE's partner in the CFM International joint venture, is the anchor in Europe, with a 25 percent share of the program. It also made sense to include Italy's Fiat because of its long-standing relationship with GE and expertise in several specific engine components. Because of the long, ongoing relationship with IHI, GE decided to approach it about participation in the new program. IHI holds an 8 percent share in the program, the same as Fiat. Participation in developing future derivatives is an opportunity available to the partners.

Up to this point, GE's colloboration with IHI had not extended beyond manufacturing. With the GE90, each partner is responsible for designing and developing its specific part of the engine. Snecma has designed and will build the compressor. IHI is responsible for several stages of turbine disks for the low-pressure turbine, the blades in those disks, and the long shaft that goes between the low-pressure turbine and the fan. IHI's interests and the ultimate content of its work share are close to what GE envisioned when approaching IHI at the outset.

In addition to design and development responsibility, program participation requires partners to make considerable capital investments in testing and manufacturing infrastructure. Because of the size and airflow of the GE90, huge new test cells are required. IHI has proceeded to make the necessary investment to build a test cell (Snecma has also built a GE90 test cell, while GE itself has built two). In addition, substantial tooling investments were also necessary to accommodate parts with the large diameter of the GE90. Partners were prepared to make these investments because of the future potential of the product.

The partnership extends for the life of the program. All commitments are measured in dollars. Typically, in a program relationship of this type, partners earn money in one of two ways. First, they may be reimbursed for their work

Annual Volume by Engine Manufacturer
Engine Orders

GE/CFMI   644
P&W/IAE   614

U.S.   1,258
non-U.S.   178

TOTAL   1,436

Number of Engines Required

■ GE/CFMI   □ P&W/IAE   □ Non-U.S.

Year

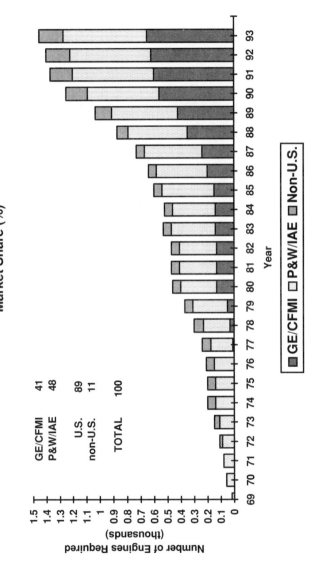

FIGURE C-2 Japan: engine orders. SOURCE: GE Aircraft Engines.

share at a fixed price, so that profits or losses are made to the extent that actual costs fall below or above the fixed price; alternatively, partners may gain a share of net revenues on the sale of engines proportional to their program share.

In the actual agreement, various protocols and rules set out partner responsibilities fairly specifically. For example, if the Federal Aviation Administration needs to extend flight tests or if there are other unanticipated costs that extend to the entire engine, the partners share these costs. If, however, there is a problem with a specific part of the engine, fixing that problem is the responsibility of the partner that designed and built the part. Generally, international partners make an up-front investment at the outset of the program in recognition of the unique contributions of the principal partner in the areas of marketing and support infrastructure, and of its established reputation in the industry. Finally, although the partners may have no formal role in marketing the engine, they do participate in support of sales campaigns in certain cases.

The GE90 is currently undergoing testing and certification; it is scheduled to enter service in 1995. Although it is not possible to assess the bottom-line impacts on the participants, GE is pleased with the partnership and with IHI's contribution to this point. The disks and turbine blades were impeccably designed and manufactured the first time around. GE has also learned some useful lessons from IHI. For example, IHI developed the casting method for the high aspect ratio turbine blades. GE gave IHI the aerodynamic coordinates on tape, which IHI quickly translated into tooling. At that point, the attachment of the blade to the disk or the tip shroud had not yet been designed. IHI said that since the major technical challenge would be to develop a good casting of the airfoil, knowing the specifics of the attachment and tip shroud was unnecessary. IHI delivered the casting in six weeks, with lumps of metal at each end that could be machined later. This fast prototyping provided insight to a "best practice" that has broad application. Under GE's old process, which had involved special casting drawings and required numerous signatures and procedures to approve changes or finalize the design, it would take a year to build tooling and prove out the casting process. In examining its process, GE realized that it was encumbered by procedures necessary for military engines, carried over to the commercial side. Making use of best practices, GE is reviewing and changing its processes for commercial engines to reduce the product development cycle time.

The origin of IHI's blade casting capability is worth noting. IHI had been developing structural and airfoil casting capability throughout its domestic network. A major advancement was realized in 1978 when as part of the F100 license agreement IHI acquired the right to cast the low-pressure turbine airfoils of equiaxed material. At that time, it was refused rights to the directionally solidified high-pressure turbine foils. In 1983 IHI acquired the rights to directionally solidified processes and materials. With this and the aid of the Technology Research and Development Institute (TRDI), IHI continued casting development and began to produce monocrystal airfoils with limited success. In 1988, the United States Air Force agreed to let IHI procure monocrystal material to use in its casting process. IHI is currently in production with airfoils of their own monocrystal process using procured material. Although the most advanced monocrystal process is not used for the GE90 blades, they are nonetheless very challenging to make, and GE rates IHI's process very highly.

F110

After a fierce competition, GE's F110 engine was selected over Pratt & Whitney's F100 as the engine for the FS-X. From the Japanese standpoint, the two major considerations were probably the higher gross weight of the FS-X aircraft and the thrust growth potential of the F110 engine.

Development work is currently going forward between GE and IHI. This involves integrating the engine with the FS-X airframe and developing the installation features. Since the engine will not be markedly different from the F110 used on the F-16 fighter, the development phase is a relatively simple process and will not involve a great deal of technology transfer or new technology development. The Japanese aim to build as much of the engine as possible from the outset, but U.S.-Japan negotiations on an FS-X production memorandum of understanding (MOU) have not been completed as of this writing. In the case of the F100, Japanese production under license has eventually reached about 75 percent.[2]

Other Collaboration

GE and IHI collaborate in several other areas. The HYPR program is covered below. In addition, in July 1992 the two companies signed a broad MOU to develop selected technologies jointly. GE initiated the MOU because it realized that opportunities to learn from IHI will increasingly arise as IHI develops its own technologies through independent efforts and as part of Japanese government-sponsored programs. GE would provide some of its know-how in exchange. The MOU provides an umbrella structure for identifying and pursuing specific opportunities. As of this writing these opportunities are under discussion, but no specific initiatives have been formalized.

GE's formal technology transfer procedures are followed on each specific program undertaken with IHI (or any other partner). First, the business unit

[2]In May 1993, the Japanese press reported that GE was willing to allow IHI to produce "most" of the F110 under license from the outset. The report also stated that GE's relatively open stance could have implications for other aspects of the program. Since the two countries have agreed on a 40 percent work share for U.S. companies and it was assumed that the United States would begin by producing a large share of the engine, an unexpectedly large Japanese share of the engine means that the U.S. share of the rest of the aircraft would have to be increased accordingly. "Nihon ni Gijutsu Kyokyu--Bei GE, Ishi-Hari ni Raisensu" (Technology to Be Provided to Japan—U.S. GE Will License to IHI), *Nihon Keizai Shimbun*, May 20, 1993, pp. 1, 11.

that wishes to transfer technology applies to a senior management technology council, which approves or disapproves specific transfers in light of the overall strategic position of GE Aircraft Engines. If the technology transfer is approved at this level, GE submits an application to the Department of State for an export license, and then to the Department of Defense (DOD) and the Department of Commerce as necessary. GE's licensed production contracts with IHI—going back to the J47—include flowback provisions in which GE will obtain improvements that IHI makes in its technology. In addition, GE rotates engineers through Japan and IHI in order to keep abreast of the Japanese partner's manufacturing and technological capabilities, as well as to manage collaborative programs. Where possible, GE uses engineers with Japanese language ability and provides language training for its employees stationed in Japan.

In addition to the technology development program, the two companies collaborate extensively on derivative engines for marine and industrial use. The LM2500, a derivative of the CF6, is used in Aegis-class cruisers and frigates, including the MSDF fleet. The engine is also used in electrical cogeneration. The GE division that makes power systems, of course, has its own extensive business and collaborative interests in Japan. However, conventional power systems take up to seven years to plan and complete. A cogeneration package using an aero-derivative engine can be put on line in about a year. IHI helps to manufacture and market these systems. For the LM2500 and the more recent LM5000—a derivative of the CF6-50—IHI has played a significant role in developing the product and aggressively marketing cogeneration systems.

PRATT & WHITNEY

Pratt & Whitney's (P&W) technology linkages with Japan are also extensive, and have included a slightly wider range of mechanisms and partners than GE's. P&W established a relationship with MHI in the 1930s that was interrupted by World War II, and has also linked with IHI and KHI. P&W's motivations for establishing technology linkages with Japan are similar to GE's— risk-sharing Japanese partners provide leverage for development funds; market access; a commitment to high quality, low-cost, and timely delivery; and increasingly new technology. Thus far, the cost and risk-sharing benefits have been most prominent. Although P&W closely monitors the technological capabilities of its Japanese partners—particularly in the manufacturing and materials areas—it has not incorporated Japanese developments to the extent that GE appears to be.

Manufacturing Alliances

Pratt & Whitney has undertaken a number of collaborative manufacturing ventures with Japanese partners over the years. These programs have covered both commercial and military engines, and have involved licensed production, risk-sharing partnerships, long-term sourcing agreements, and subcontracted production.

In 1978, the F100 engine was selected as the engine to be used on Japan's F-15s. The first two complete engines—to be used by IHI in calibrating its testing equipment—were delivered in May 1979, and eight knock-down kits were delivered beginning in August 1980. In September 1981, IHI produced the first engine under license, and 290 F100-IHI-100 engines were made under this agreement through April 1990. Some of the materials and the electronic engine controls were held back by DOD, but IHI manufactures about 75 percent of the engine by dollar value.

Over the last several years, the F100 relationship has evolved further, as IHI incorporates improvements that P&W developed for the U.S. version of the F100. From April 1990 until September 1993, IHI produced under license the F100-IHI-100BJ—which incorporated an increased life core—at the rate of two engines per month. In September 1993, IHI began production of the IDEEC F100-PW-220E engine at the rate of two per month, and it was scheduled to begin retrofitting prior engines with 220E hardware at the rate of three per month in March 1994. The major advance in the 220E is digital electronic control. In all, IHI is scheduled to produce 472 F100s of all versions under the current contract.

P&W launched an earlier and less extensive military licensed production agreement in 1971 with MHI covering the JT8D-9 engine. MHI produced about 70 of the engines over 10 years for Japan's C-1 military transport. In 1984, MHI became a 2.8 percent risk-sharing partner in the manufacture of a derivative product, the 20,000-pound JT8D-200, which powers the McDonnell Douglas MD-80 series. Under this agreement, MHI is responsible for the manufacture of various turbine blades, disks, and cases. In joining an existing program, MHI had no development role.

P&W has two Japanese partners in the PW4000 program, a large engine that powers some versions of the Boeing 767 and whose derivatives will be carried on some versions of the Boeing 777. The engine was originally developed in the late 1970s. Kawasaki became a 1 percent risk-sharing partner in 1985, and it has continued at that level since then. It is responsible for manufacturing several airseals, a shaft coupling, and a pump. MHI signed on as a 1 percent risk-sharing partner in the PW4000 program in 1989, and since then its participation grew to 5 percent in 1991 and 10 percent in 1993. MHI is responsible for manufacturing various turbine blades and vanes, turbine and compressor disks, active clearance control components, and combustion chambers. Beginning participation at such a low level reflected P&W's desire to "test the waters" and establish a working relationship with its partners before investing a great deal in the alliance. The increase in MHI's share since 1989 has come about as a result of mutual satisfaction with the relationship and desire to expand it.

In addition to risk-sharing agreements with MHI and KHI in commercial engines, P&W has a long-term sourcing agreement with IHI to produce the big shaft connecting the high- and low-pressure turbines for the JT9D, PW2000, and PW4000. P&W has also subcontracted to MHI production of components for the JT8D (low-pressure turbine and low-pressure compressor disks) and JT9D (low-pressure turbine blades).

IHI now manufactures all of Pratt & Whitney's long shafts. Utilizing and improving upon the process transferred in connection with the F100 program, IHI has become a world-class center for the production of long shafts of more than 8 feet. As mentioned earlier, IHI will be manufacturing the long shaft for the GE90, and it manufactures all of Rolls Royce's shafts as well. This specialization is not uncommon in the engine business—Fiat dominates the manufacture of gear boxes, and Volvo is strong in casings. Although IHI's dominance in shafts raises issues of dependence and possible supply disruption, the engine "primes" manage this dependence by maintaining some capability of their own. It is also widely believed that any attempt by a supplier of critical engine components to use delay or denial to extract money or technology from the primes would spell death for that supplier in the international market. The focused manufacturing approach does carry significant benefits in terms of cost and quality.

INTERNATIONAL AERO ENGINES (IAE)

International Aero Engines is a global consortium that developed and is currently manufacturing and marketing the V2500 engine. It consists of Pratt & Whitney, Rolls Royce, Fiat, MTU, and Japan Aero Engines Company (JAEC).[3] Although IAE currently has just one product—the V2500 engine—the alliance includes a 30-year commitment to produce engines in the 18,000 to 30,000-pound range and has provisions for studies of engines up to 35,000 pounds of thrust. As of this writing, 104 V2500s have been delivered, the order backlog stands at 284, and airlines hold options on 302 more.

The partner companies in IAE were responsible for developing as well as building their share of the engine. The lead partners—Pratt & Whitney and Rolls Royce—both hold 30 percent shares in the program. P&W is responsible for the high-pressure turbine and the combustion system, whereas Rolls Royce designed and manufactures the high-pressure compressor and the lubrication system. In addition to the program shares, P&W holds a separate contract for overall engine management and manages the electronic engine control. Rolls Royce manages the design and manufacture of the nacelle and is also responsi-

---

[3]Background on the formation of IAE is contained in David C. Mowery, *Alliance Politics and Economics: Multinational Joint Ventures in Commercial Aircraft* (Cambridge, Mass.: Ballinger Publishing Company, 1987) and Richard J. Samuels, *Rich Nation, Strong Army: National Security, Ideology, and the Transformation of Japan* (Ithaca, N.Y.: Cornell University Press, forthcoming 1994).

ble for V2500 training activities for airlines. Germany's MTU holds 11 percent, and builds the low-pressure turbine, whereas Fiat holds 6 percent and is responsible for the accessory gear box and turbine exhaust case. For both MTU and Fiat, V2500 responsibilities are similar to their participation in PW2037 development and manufacturing. JAEC holds 23 percent of IAE, and is itself a joint venture of IHI (with 60 percent of JAEC), Kawasaki (25 percent), and MHI (15 percent). JAEC is responsible for the fan and the low-pressure compressor. Although JAEC has representatives in the marketing department of IAE, P&W and Rolls Royce are fundamentally responsible for marketing. Technical support at the airlines is accomplished largely through P&W's existing system.

IAE and the V2500 program were carefully structured to minimize technology transfer between the partners. This was partly motivated by DOD concerns about transferring P&W's high-pressure turbine technologies, but it also reflects the competitive concerns of the partners.[4] Like the CFM56, the GE90, and other collaboratively designed engine programs, the V2500 utilizes a modular design in which a complete engine can be assembled and tested without a great deal of knowledge exchange concerning the individual pieces. The benefits of risk and cost sharing, specialized manufacturing, and market leverage must be balanced against the built-in overhead cost and time disadvantages of involving so many companies, as well as the extra time and care required to negotiate interface designs that limit the flow of technology. Still, Pratt & Whitney will benefit to the extent that there are generic rules and practices arising from the V2500 experience that can be applied to managing future collaborative programs.

All of the non-U.S. members of IAE received support from their governments for their participation. Rolls Royce received a $150 million no-interest loan from the British government, slightly less than half of the cost of participation that it estimated at the outset, to be repaid through a royalty on each sale.[5] JAEC has received annual payments of $20 million to $25 million from the Ministry of International Trade and Industry (MITI) since the start of the abortive FJR710 program in the early 1970s, and this support has continued through V2500 development, covering roughly 75 percent of JAEC's development costs, 66 percent of testing costs, and 50 percent of the production tooling and nonrecurring startup costs.[6] Repayment with interest of these success-conditional loans is slated to commence when the program breaks even. Exact figures for government support extended to MTU and Fiat are more difficult to obtain.[7]

The V2500 faces tough competition from the CFM International CFM56, but appears to be gaining greater market acceptance over time. Although the formal IAE agreement requires the partners to work together on engines in the

---

[4]Mowery, ibid., p. 94.
[5]Ibid., p. 93.
[6]Ibid.
[7]Ibid., p. 94.

18,000 to 30,000-pound range, the wording of the agreement is very complex and thrust is not the only determining requirement. One of the key aspects for companies in forming joint ventures and alliances is defining the product scope in a way that the partnership can be expanded if desirable from a business standpoint, but does not constrain the partners as they pursue their individual strategies.

HYPR AND OTHER JAPANESE GOVERNMENT PROGRAMS

The Japanese Supersonic/Hypersonic Propulsion Technology Program (JSPTP or HYPR), was launched by MITI in 1989, with funding originally set at $200 million over eight years. It is now expected that the program will be stretched to ten years. The ultimate goal of the program is the development of a scale prototype turbo-ramjet, Mach 5 methane-fueled engine. The program is administered by MITI through its Agency of Industrial Science and Technology and the quasi-governmental New Energy Development Organization. The specification of a Mach 5 methane engine was partially determined by Japanese bureaucratic politics. MITI was not able to obtain Ministry of Finance approval to fund a supersonic engine program, but it could utilize funds earmarked for energy conservation R&D if the targeted development utilized an alternative fuel such as methane.

The Japanese partners—IHI, Kawasaki, and MHI—receive 75 percent of the funding and take the lead on technology development and design. HYPR is significant in that it is the first of Japan's national R&D projects to contemplate international participation from the outset as an integral feature of the program. The foreign participants—who receive 25 percent of the funding—are Pratt & Whitney, GE, Rolls Royce, and Snecma. The formal agreement between MITI and the foreign engine companies was signed in early 1991. The process of negotiating the participation of the foreign engine companies was somewhat long and complex. The major stumbling block arose surrounding the treatment of intellectual property generated in the project. The standard treatment of intellectual property in Japan's government-sponsored R&D is that the gov-ernment owns 50 percent and exercises effective control over the disposition of intellectual property rights (IPR). The four foreign companies, wanting to avoid possible future restrictions on IPR, joined together to negotiate with MITI as a united front. This process led to an agreement and a change in Japan's laws governing the administration of government-sponsored R&D. Purely domestic projects follow the same rules as always, but IPR is treated differently in desig-nated international projects such as HYPR as a result of the change. The fore-ground results and patents of technology developed in the program are owned jointly by the foreign and Japanese companies that developed them. Individual companies can use their own results without restriction, but they must negotiate with MITI over fees if an outward license is contemplated. Access to patents and technical information arising in other parts of the program is open, and technology can be licensed. The background technology that individual companies bring to the project is controlled by the owner.

The Japanese companies are taking the lead on various program elements. To this point, the participating U.S. companies report satisfaction with the progress of the work and the flow of information. From the point of view of GE and Pratt & Whitney, the main motivation for participating is that taking a leadership role in the Japanese program is preferable to a major supersonic/hypersonic engine program going forward without U.S. involvement. By participating, GE and Pratt & Whitney gain insights into the basic design decisions and capabilities of the Japanese members of HYPR. Besides, because of MITI funding, participation is not costly for the foreign firms. It was necessary for the two companies to touch bases with the Departments of Defense and Commerce at the outset, and to convince them of the rationale. Eventually, the U.S. government was persuaded that "riding in front of the stampede" made more sense than sitting on the sidelines. U.S. engine makers believe that as a major terminus for flights of the next-generation supersonic transport, Japan will inevitably be involved in its development. GE and Pratt & Whitney are collaborating on research funded by the National Aeronautics and Space Administration (NASA) on high-speed civil transport propulsion targeting an engine in the Mach 2 to 2.5 range. The NASA program involves a much higher funding level than Japanese government support of HYPR. The U.S. engine makers are not transferring technology from this work to the Japanese.

The basic interaction between foreign and Japanese companies in HYPR is participation in design review and analysis in designated program areas. GE or Pratt & Whitney will look at the designs of, respectively, IHI and KHI, critique the work, and coach them on possible new directions. Each of the foreign companies assigns five to ten engineers to the project. On the Japanese side, the HYPR management headquarters has a staff of 11, but a minimum of 500 engineers in the three companies charge at least part-time to the program.[8]

Since the program is currently in its fourth year and will probably run for ten, the impacts and implications cannot be assessed precisely. The eventual impact will depend a great deal on the timing and mechanism for developing propulsion for the next-generation supersonic transport. While foreign participation allows the major international players to gain knowledge about Japan's approach, the Japanese participants gain design insights from foreign coaching. Also, international participation in HYPR has itself served to give credibility to Japanese efforts to play a significant role in international advanced engine programs and to other Japanese government efforts to organize international R&D collaboration.

---

[8]"International Group to Build Combined Cycle Hypersonic Engine," *Aviation Week & Space Technology*, August 17, 1992, p. 50.

The Japanese government also funds several other programs that have implications for future aircraft propulsion systems. The one that is most closely linked to HYPR is the research program on high-performance materials organized under MITI's "Jisedai" or Next-Generation Technology Development funding pool. The program, which began in 1989 and is scheduled to run through 1996, is organized into two parts. One branch includes six government laboratories, while the other—which is known as the Research and Development Institute of Metals and Composites for Future Industries (RIMCOF) is composed of nine companies and four universities. RIMCOF itself was launched in 1981, and completed two MITI-sponsored R&D programs on composites and crystalline alloys from 1981 to 1988. Toray was the main industrial participant and beneficiary of the composites project.[9] In the current project, the focus is on intermetallic compounds, heat-resistant fibers, composites, and reinforced intermetallic composites that could be utilized in supersonic or hypersonic engines.

Japanese government and industry have been working together on subsonic engine technologies as well. The Frontier Aircraft Basic Research Center Co., Ltd. (FARC) was established by the Key Technology Center in 1986 to develop the technology required for an advanced turboprop engine. FARC, which operated through the beginning of 1993, included 34 companies in all. In addition to the major engine and airframe "heavies," auto, materials, and machinery manufacturers are also involved.[10]

In addition to these ongoing programs, the Japanese government—mainly MITI and TRDI—are conducting a number of feasibility studies aimed at significantly upgrading Japan's engine testing facilities over the next decade. The most important of these is an altitude test facility to be built in Hokkaido at a projected total cost of $140 million.

JAPANESE CAPABILITIES IN THE AERO ENGINE BUSINESS

Japanese aircraft engine makers have effectively leveraged private and public resources in international alliances and public R&D projects to improve and deepen their technological and manufacturing capabilities. Individually or as a group, Japanese companies are well positioned to continue to participate in international engine development programs at increased levels of technical and manufacturing responsibility. Japan's government technology programs and corporate strategies are aimed at gaining world leadership in some aspects of propulsion materials and other critical technologies.

As in the aircraft systems segment, barriers to Japan's entry as a major player at the level of today's international engine "big three" remain. To begin

[9]Michael Dornheim, "MITI Pursues Improved High Temperature Materials," *Aviation Week and Space Technology*, August 17, 1992, p. 54.
[10]See Samuels, op. cit., for a detailed description of FARC.

with, industry and government have not yet shown a willingness to invest the resources necessary to enter the market at that level. As in aircraft integration, this would require either a series of large, risky indigenous projects to establish technical and market credibility, or an acquisition. In the case of the American companies, DOD has guarded hot section technologies over the years. While it is currently difficult to conceive of a circumstance in which DOD would allow the transfer of these technologies through license or acquisition, the U.S. policy context is changing.

Japan's current technological capabilities are quite impressive in several areas of engine manufacturing and technolgy. The Japanese can manufacture most parts of a modern engine and can design key pieces. IHI in particular is quite strong in application of technology once it has mastered the basic concept. Its manufacturing practices—including total preventive maintenance—are very effective, as are its laser drilling capabilities. The proof of this is in the product—IHI and the other Japanese companies hold tolerances very well.

The Japanese engine makers do have significant weaknesses. Across the board, the Japanese companies are weak in software and lack sophistication in the analytical tools necessary to do world-class design. For example, when asked to design a compressor blade, the Japanese are capable of very competent mechanical design. However, it takes time for them to experiment and trade off the mechanical and aerodynamic features. The U.S. engine companies have computer programs that can optimize both mechanical and aerodynamic characteristics in designing blades.

The Japanese are aware of their weaknesses in software and systems integration methodology, and are asking more often for access to analytical tools in their international alliances. These are the technological "crown jewels" that the U.S. engine companies guard fiercely. Even if they were willing to transfer them, many of the the management methodologies are best learned by actually doing a complete engine program. In the case of some design tools, such as CATIA, the Japanese may possess the software, but they have only a thin experimental data base to plug into it to gain optimum value from the software. Finally, the Japanese engine makers have relatively high unit manufacturing costs and overhead, disadvantages that are currently being exacerbated by the latest yen appreciation.

MITI and the JDA have supported the Japanese aircraft and engine industries with the aim of helping them to become world leaders. There may be a certain amount of frustration that industry is not further along, given the significant amount of public funds spent on various aspects of aircraft development. There has been a recent willingness to allow or even encourage nontraditional Japanese players to test the waters. In the engine world, these are the largest auto companies—Toyota, Nissan, and Honda.

Despite these weaknesses, the Japanese have developed a significant manufacturing and technological base in the engine business. Government and industry continue to team in the development of advanced technologies—in materials and other areas. The HYPR project illustrates a creative approach to international collaboration and reflects the long-term orientation of Japanese strategy making. While the global leaders in the industry are pursuing breakthrough new products, Japanese participation in international engine programs has increased over the past decade, and Japanese government-sponsored programs are aimed at developing a technology base to further expand this role. As in airframes, new directions in international collaboration—either with the Russians or other new partners, or through selective utilization of experts idled by worldwide defense cuts—are feasible strategies. In parallel with the airframe business, current global restructuring is challenging the Japanese as it is challenging other players. However, the rewards are likely to go to companies and nations committed for the long haul, as the Japanese clearly are.